T0137250

Question Answering over Text and Knowledge Base

Saeedeh Momtazi • Zahra Abbasiantaeb

Question Answering over Text and Knowledge Base

Saeedeh Momtazi
Computer Engineering Department
Amirkabir University of Technology
Teheran, Iran

Zahra Abbasiantaeb
Computer Engineering Department
Amirkabir University of Technology
Teheran, Iran

ISBN 978-3-031-16554-2 ISBN 978-3-031-16552-8 (eBook)
https://doi.org/10.1007/978-3-031-16552-8

This Springer imprint is published by the registered company Springer Nature Switzerland AG
The registered company address is: Gewerbestrasse 11, 6330 Cham, Switzerland

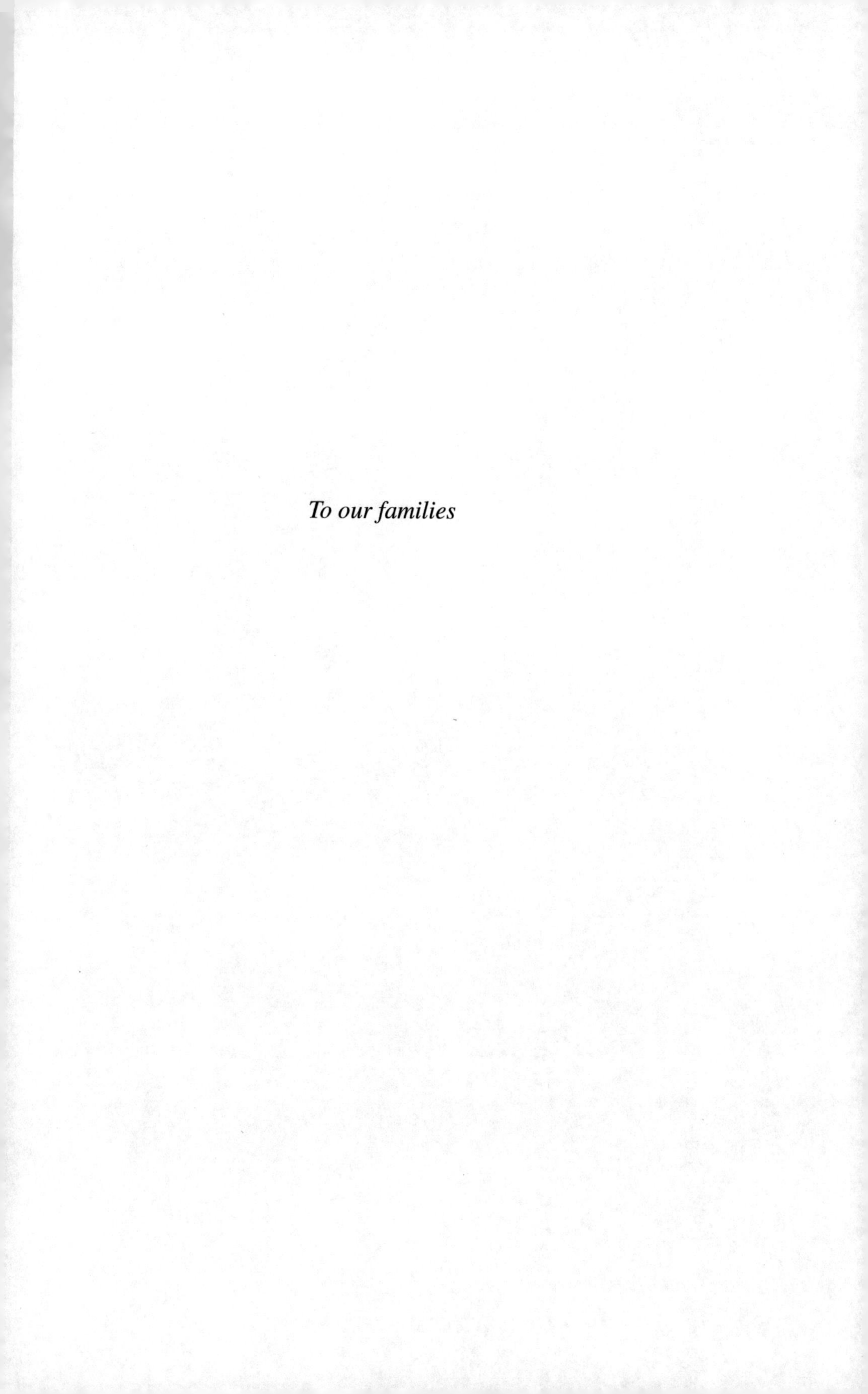

To our families

Preface

The advance of natural language processing and information retrieval models has significantly influenced the field of question answering systems. There have been an increased attention and research on question answering systems. By the advent of renown question answering systems like Siri and Google Assistant, the importance of the question answering system in this era is evident.

The importance of question answering systems in the fields of natural language processing and information retrieval, existence of a large number of scientific papers, and advances in this field from basic information retrieval models to deep neural networks convinced us to write this book. Moreover, the available survey papers on question answering systems are not comprehensive, and they only cover the studies on one type of question answering systems. As a result, they are not enough for obtaining a big picture of question answering systems advances and categories. Hence, publishing a specific and comprehensive book on question answering systems that can be used by researchers starting in this field seems necessary. Such a book can help the readers to follow the state-of-the-art research by providing essential and basic knowledge.

Consequently, we have sensed the need to publish a book on question answering systems and started writing this book. Both authors of the book have several years of experience and research on question answering systems. Saeedeh Momtazi completed her PhD dissertation on question answering systems, and after that she ran a couple of tutorials on the field and later on as an associate professor supervised different projects on question answering. Zahra Abbasiantaeb completed her MSc on question answering systems and followed this topic through publishing surveys and technical papers.

In summary, we have done our best to convey the big picture to the readers and also cover as much detail as possible essential for following the latest papers. We have studied the question answering systems in three main categories based on the source of the data that can be unstructured text, structured knowledge graphs, or a combination of both. In addition, we provided the required materials for understanding the technical chapters and reviewed models in the book.

Tehran, Iran
May 2022

Saeedeh Momtazi
Zahra Abbasiantaeb

Acknowledgements

No worthwhile work was ever completed in isolation. Finishing this book could not have been possible without the help of several people. We owe them a debt of gratitude for all their support and kindness.

First of all, our appreciation goes to Ralf Gerstner, executive editor for Computer Science at Springer, for all his help and support during writing and revision of the proposal and the book.

We are very grateful to the Computer Engineering Department of Amirkabir University of Technology for providing a pleasant work environment, where we could devote a substantial amount of time to writing this book.

We are very thankful for Prof. Dr. Dietrich Klakow and Prof. Dr. Felix Naumann who had substantial influence in our research journey on question answering systems and education.

Last but not least, appreciation goes to our families for their true love. Their unconditional support and care were the best help. We are grateful for all their patience and tolerance in difficult conditions.

Contents

Chapter 1
Introduction

Abstract This book aims at providing a coherent and complete overview of the presented question answering (QA) approaches. The QA models are studied from different perspectives. The available models for QA are divided into three major categories, namely, TextQA, KBQA, and hybrid QA. This chapter describes the importance of QA systems compared to the search engines and discusses how QA systems can provide the required information for users by short, concrete answer strings. A brief explanation of two main categories of QA systems, namely, TextQA and QA over knowledge, and a detailed comparison between these two categories will be provided.

1.1 Introduction

Most of the computer users find their required information by surfing the Web and searching through Internet pages. By the existence of several free search engines, such as Google, Bing, and Yahoo, Internet searching is the first and easiest way of providing the required information for users. Users type their intended information as a small set of keywords and receive a list of relevant documents. The returned list of relevant documents, however, is not enough for users to find the exact information that they are looking for. In addition, the users have to spend a lot of time to manually retrieve the exact information they need from a large number of retrieved documents.

QA systems are one of the effective tools for users who want to receive an exact answer to their questions rather than being overwhelmed with a large number of retrieved documents, which they must then sort through to find the desired answer.

In contrast to search engines which return a list of retrieved documents, the goal of QA systems is to return a short answer string which exactly addresses the user's question. In addition to the different formats in the output of QA systems compared to search engines, they also have a different form of input. While search engines need a set of keywords as input, the input of QA systems is a natural language question.

Having such differences between search engines and QA systems indicates that developing a QA system needs more efforts and studies beyond search

S. Momtazi, Z. Abbasiantaeb, *Question Answering over Text and Knowledge Base*,
https://doi.org/10.1007/978-3-031-16552-8_1

engine development. This goal is usually achieved using a combination of various important techniques, including natural language processing, information retrieval, information extraction, knowledge graph processing, machine learning, and deep learning.

QA systems can be divided into different types from different perspectives (Dimitrakis et al., 2020). We discuss the available perspectives in the following.

- **Domain of information:** Based on this perspective, the QA systems are divided into two categories, namely, open domain and close domain. In the open-domain systems, the subject of the questions is not restricted, and the system can answer every question regardless of its domain. The closed-domain systems are developed for answering questions of a specific domain, e.g., medical, movies, hotels, and museums. Domain of questions influences the development of models. The main influential factor is the size of the available data. The size of the data in open-domain QA is much larger than the size of the data in the closed-domain QA. The data size influences the choice of machine learning and deep learning models, because deep neural networks require a large size of data for training. Another influential factor is the context of questions. For example, in the case of closed-domain QA, solving the problem of word sense disambiguation is much easier because of the clarity context.
- **Question type:** The questions are divided into different categories based on the required technique for answering them, perplexity, and type of their answer. The categories are as follows:

  1. Factoid: The factoid questions usually start with "when," "where," and "who" words, and they require a single fact from knowledge base or a small piece of text to be answered.
  2. Yes/No (Confirmation): The yes/no questions are answered with yes or no.
  3. Definition: Definition of a term is required for answering these questions.
  4. Causal: These questions usually start with "how," "why," and "what" words and are answered with one or more consequences of a fact.
  5. Procedural: They are answered with a sequence of actions that must be performed to complete a task.
  6. Comparative: Answer to these questions is the comparison of several subjects from different perspectives.
  7. With Examples: These questions aim to find examples for explaining what the question asks for.
  8. Opinion: These questions ask about opinion of a person on a specific subject.

- **Knowledge source:** The QA systems are divided into three categories according to the type of the data source that they use for retrieving answers including textual data, knowledge bases, and combination of them. The approaches and techniques that are used in each category are different, and their overall architectures are different as a result.

1.2 Book Overview

QA techniques belong to two major paradigms based on the used dataset. The first paradigm is QA over text (TextQA), and the second paradigm is QA over knowledge base (KBQA). Researches on each of these paradigms belong to two major approaches, namely, traditional approaches and deep learning models. One chapter is devoted to each of these paradigms, and a broad explanation about architecture or statistics of the proposed models is provided. We also cover the systems that benefit from both approaches within a combined architecture.

TextQA uses textual data and KBQA uses a knowledge base for extracting the answer of a given question. In the overall architecture of the TextQA approach, a set of documents are available as data source, and given an input question, the answer of the question is retrieved from documents. In the TextQA architecture, through several steps, the documents are processed to extract the relevant documents. In this way, the search space is limited to a specific number of documents, and a set of candidate answers that are potential answers of the input question are selected in the answer retrieval step. In the next steps, the answer sentence which carries the exact answer of the question is selected, and the answer keyword is extracted.

Most of the current research studies in this area focus on the answer sentence selection task in overall architecture. We define the task of TextQA as follows: given an input question and a list of candidate answers, the system must rank the list of candidate answers based on the probability of being a correct answer. A large number of traditional and neural models are proposed for calculating the probability of being a correct answer for each of the candidate answers given the question. The system can then classify candidate answers as correct or incorrect.

We study the available traditional (non-neural) models for TextQA in Sect. 5.2. The neural models calculate the explained probability by measuring the semantic similarity or matching score of the question with each candidate answer sentence.

The available neural models follow a text matching structure. The available models are divided into three categories based on the architecture including representation-based (5.3.1), interaction-based (5.3.2), and hybrid (5.3.3). The representation-based methods (Severyn & Moschitti, 2015; Tan et al., 2016; Tay et al., 2017; Wang & Nyberg, 2015; Yin et al., 2016b; Yu et al., 2014) model the semantic representation of each sentence separately and then by comparing the semantic representations calculate the matching score of the question and each candidate answer. The interaction-based methods (Garg et al., 2020; Wan et al., 2016b) form an interaction matrix by modeling the interaction of two input text (question and candidate answer sentence) by comparing each term from both texts. The matching score is calculated based on the interaction matrix. The hybrid method (Bian et al., 2017; He & Lin, 2016b; Tan et al., 2016; Tay et al., 2018; Wan et al., 2016a; Wang & Jiang, 2017; Wang et al., 2016, 2017; Yang et al., 2019; Yin et al., 2016b; Yoon et al., 2019) is a combination of both the interaction-based and representation-based models.

In KBQA, the source of data for extracting answers is knowledge base. A knowledge graph includes entities connected to each other with edges denoted by relation. Information is stored in the form of facts, and each fact is a triple indicating the relation (or predicate) between two entities: [head_entity, predicate, tail_entity].

Considering the type of input question, KBQA is divided into two types, namely, simple questions and complex questions. A simple question is defined as a question which can be answered by one fact from a knowledge graph, while a complex question requires reasoning over a set of facts to retrieve the answer. For example, the input question *"where was Einstein born?"* is a simple question as answer of the question is stored in one fact like ["Einstein," "place_of_birth," "ULM"], while the input question *"where was the inventor of automobile born?"* is a complex question because the name of the inventor of the electricity must be retrieved from one fact ["Karl Benz," "Inventor_of," "Automobile"] and then his place of birth must be retrieved from another fact ["Karl Benz," "place_of_birth," "Mühlburg"].

We studied the available models for KBQA in two separate categories. The models proposed for simple questions (He & Golub, 2016a; Huang et al., 2019; Mohammed et al., 2018; Sorokin & Gurevych, 2017; Yin et al., 2016a,c) are studied in Sect. 6.3, and the models proposed for complex questions (Hao et al., 2019; Vakulenko et al., 2019; Zafar et al., 2018; Zhu et al., 2020) are studied in Sect. 6.4.

In Chap. 2, we present the history of QA systems and the architecture of different QA approaches. The chapter starts with early closed-domain QA systems, including BASEBALL (Green et al., 1963) and LUNAR (Woods, 1977), and reviews different generations of QA until state-of-the-art hybrid models.

Chapter 3 is devoted to explaining the datasets and the metrics used for evaluating TextQA and KBQA.

Chapter 4 introduces the neural and deep learning models used in QA systems. This chapter includes the required knowledge of deep learning and neural text representation models for comprehending the QA models over text and QA models over knowledge base explained in Chaps. 5 and 6, respectively.

In some of the KBQA models, the textual data is also used as another source besides the knowledge base (Savenkov & Agichtein, 2016; Sun et al., 2018; Xu et al., 2016). We study these models in Chap. 7 as "KBQA enhanced with textual data."

In Chap. 8, a detailed explanation of some well-known real applications of the QA systems is provided. Eventually, open issues and future work on QA are discussed in Chap. 9.

The materials of this book are gathered from papers published by renowned publication in the field of QA. We have selected the materials to be included in the book based on several criteria, e.g.:

1. Ensuring the depth and cover of the materials on each topic: To this aim, we have selected the papers which introduce a new idea for solving the problem. In addition, we tried to include papers which are empirically effective.
2. Including the recent and novel models: We have done our best to cover the state-of-the-art and recent works on the topic.

This document covers topics on both QA over text and QA over knowledge base. But there are other dimensions of QA which have not been covered in this book, including reading comprehension systems and cognitive models in QA.

References

Bian, W., Li, S., Yang, Z., Chen, G., & Lin, Z. (2017). A compare-aggregate model with dynamic-clip attention for answer selection. In *Proceedings of the 2017 ACM on Conference on Information and Knowledge Management, CIKM '17*, New York, NY, USA (pp. 1987–1990). ACM. ISBN:978-1-4503-4918-5. https://doi.org/10.1145/3132847.3133089.

Dimitrakis, E., Sgontzos, K., & Tzitzikas, Y. (2020). A survey on question answering systems over linked data and documents. *Journal of Intelligent Information Systems, 55*(2), 233–259.

Garg, S., Vu, T., & Moschitti, A. (2020). Tanda: Transfer and adapt pre-trained transformer models for answer sentence selection. In *Thirty-Fourth AAAI Conference on Artificial Intelligence*.

Green, B., Wolf, A., Chomsky, C., & Laughery, K. (1963). Baseball: an Automatic Question Answerer. In E. Figenbaum, & J. Fledman (Eds.), *Computers and thoughts*. McGraw-Hill.

Hao, Z., Wu, B., Wen, W., & Cai, R. (2019). A subgraph-representation-based method for answering complex questions over knowledge bases. *Neural Networks, 119*, 57–65.

He, X. & Golub, D. (2016a). Character-level question answering with attention. In *Proceedings of the 2016 Conference on Empirical Methods in Natural Language Processing*, Austin, Texas (pp. 1598–1607). Association for Computational Linguistics. https://doi.org/10.18653/v1/D16-1166. https://www.aclweb.org/anthology/D16-1166.

He, H. & Lin, J. (2016b). Pairwise word interaction modeling with deep neural networks for semantic similarity measurement. In *Proceedings of the 2016 Conference of the North American Chapter of the Association for Computational Linguistics: Human Language Technologies*, San Diego, CA (pp. 937–948). Association for Computational Linguistics. https://doi.org/10.18653/v1/N16-1108.

Huang, X., Zhang, J., Li, D., & Li, P. (2019). Knowledge graph embedding based question answering. In *Proceedings of the Twelfth ACM International Conference on Web Search and Data Mining, WSDM '19*, New York, NY, USA (pp. 105–113). Association for Computing Machinery. https://doi.org/10.1145/3289600.3290956.

Mohammed, S., Shi, P., & Lin, J. (2018). Strong baselines for simple question answering over knowledge graphs with and without neural networks. In *Proceedings of the 2018 Conference of the North American Chapter of the Association for Computational Linguistics: Human Language Technologies, Volume 2 (Short Papers)*, New Orleans, Louisiana (pp. 291–296). Association for Computational Linguistics. https://doi.org/10.18653/v1/N18-2047.

Savenkov, D., & Agichtein, E. (2016). When a knowledge base is not enough: Question answering over knowledge bases with external text data. In *Proceedings of the 39th International ACM SIGIR Conference on Research and Development in Information Retrieval, SIGIR '16*, New York, NY, USA (pp. 235–244). Association for Computing Machinery. ISBN:978-1-4503-4069-4. https://doi.org/10.1145/2911451.2911536.

Severyn, A., & Moschitti, A. (2015). Learning to rank short text pairs with convolutional deep neural networks. In *SIGIR*.

Sorokin, D. & Gurevych, I. (2017). End-to-end representation learning for question answering with weak supervision. In M. Dragoni, M. Solanki, & E. Blomqvist (Eds.), *Semantic web challenges*. Springer International Publishing.

Sun, H., Dhingra, B., Zaheer, M., Mazaitis, K., Salakhutdinov, R., & Cohen, W. (2018). Open domain question answering using early fusion of knowledge bases and text. In *Proceedings of the 2018 Conference on Empirical Methods in Natural Language Processing*, Brussels,

Belgium (pp. 4231–4242). Association for Computational Linguistics. https://doi.org/10.18653/v1/D18-1455. https://www.aclweb.org/anthology/D18-1455.
Tan, M., dos Santos, C., Xiang, B., & Zhou, B. (2016). Improved representation learning for question answer matching. In *Proceedings of the 54th Annual Meeting of the Association for Computational Linguistics (Volume 1: Long Papers)*, Berlin, Germany (pp. 464–473). Association for Computational Linguistics. https://doi.org/10.18653/v1/P16-1044.
Tay, Y., Phan, M. C., Tuan, L. A., & Hui, S. C. (2017). Learning to rank question answer pairs with holographic dual lstm architecture. In *Proceedings of the 40th International ACM SIGIR Conference on Research and Development in Information Retrieval, SIGIR '17*, New York, NY, USA (pp. 695–704). ACM. ISBN:978-1-4503-5022-8. https://doi.org/10.1145/3077136.3080790.
Tay, Y., Tuan, L. A., & Hui, S. C. (2018). Multi-cast attention networks. In *Proceedings of the 24th ACM SIGKDD International Conference on Knowledge Discovery & Data Mining, KDD '18*, New York, NY, USA (pp. 2299–2308). Association for Computing Machinery.
Vakulenko, S., Fernandez Garcia, J. D., Polleres, A., de Rijke, M., & Cochez, M. (2019). Message passing for complex question answering over knowledge graphs. In *Proceedings of the 28th ACM International Conference on Information and Knowledge Management* (pp. 1431–1440).
Wan, S., Lan, Y., Guo, J., Xu, J., Pang, L., & Cheng, X. (2016a). A deep architecture for semantic matching with multiple positional sentence representations. In *Proceedings of the Thirtieth AAAI Conference on Artificial Intelligence, AAAI'16* (pp. 2835–2841). AAAI Press.
Wan, S., Lan, Y., Xu, J., Guo, J., Pang, L., & Cheng, X. (2016b). Match-srnn: Modeling the recursive matching structure with spatial rnn. In *IJCAI*.
Wang, D. & Nyberg, E. (2015). A long short-term memory model for answer sentence selection in question answering. In *Proceedings of the 53rd Annual Meeting of the Association for Computational Linguistics and the 7th International Joint Conference on Natural Language Processing (Volume 2: Short Papers)*, Beijing, China (pp. 707–712). Association for Computational Linguistics. https://doi.org/10.3115/v1/P15-2116.
Wang, S. & Jiang, J. (2017). A compare-aggregate model for matching text sequences. In *Proceedings of the 5th International Conference on Learning Representations (ICLR)*.
Wang, B., Liu, K., & Zhao, J. (2016). Inner attention based recurrent neural networks for answer selection. In *Proceedings of the 54th Annual Meeting of the Association for Computational Linguistics (Volume 1: Long Papers)*, Berlin, Germany (pp. 1288–1297). Association for Computational Linguistics. https://doi.org/10.18653/v1/P16-1122.
Wang, Z., Hamza, W., & Florian, R. (2017). Bilateral multi-perspective matching for natural language sentences. In *Proceedings of the Twenty-Sixth International Joint Conference on Artificial Intelligence (IJCAI-17)* (pp. 4144–4150). https://doi.org/10.24963/ijcai.2017/579.
Woods, W. A. (1977). Lunar rocks in natural english: Explorations in natural language question answering. In A. Zampolli (Ed.), *Linguistic structures processing* (pp. 521–569). North-Holland.
Xu, K., Feng, Y., Huang, S., & Zhao, D. (2016). Hybrid question answering over knowledge base and free text. In *Proceedings of COLING 2016, the 26th International Conference on Computational Linguistics: Technical Papers*, Osaka, Japan (pp. 2397–2407). The COLING 2016 Organizing Committee. https://www.aclweb.org/anthology/C16-1226.
Yang, R., Zhang, J., Gao, X., Ji, F., & Chen, H. (2019). Simple and effective text matching with richer alignment features. In *Proceedings of the 57th Annual Meeting of the Association for Computational Linguistics* (pp. 4699–4709). Association for Computational Linguistics.
Yin, J., Jiang, X., Lu, Z., Shang, L., Li, H., & Li, X. (2016a). Neural generative question answering. In *Proceedings of the Workshop on Human-Computer Question Answering*, San Diego, California (pp. 36–42). Association for Computational Linguistics. https://doi.org/10.18653/v1/W16-0106. https://www.aclweb.org/anthology/W16-0106.
Yin, W., Schütze, H., Xiang, B., & Zhou, B. (2016b). Abcnn: Attention-based convolutional neural network for modeling sentence pairs. *Transactions of the Association for Computational Linguistics, 4*, 259–272.

Yin, W., Yu, M., Xiang, B., Zhou, B., & Schütze, H. (2016c). Simple question answering by attentive convolutional neural network. In *Proceedings of COLING 2016, the 26th International Conference on Computational Linguistics: Technical Papers*, Osaka, Japan (pp. 1746–1756). The COLING 2016 Organizing Committee. https://www.aclweb.org/anthology/C16-1164.

Yoon, S., Dernoncourt, F., Kim, D. S., Bui, T., & Jung, K. (2019). A compare-aggregate model with latent clustering for answer selection. In *Proceedings of the 28th ACM International Conference on Information and Knowledge Management* (pp. 2093–2096).

Yu, L., Hermann, K. M., Blunsom, P., & Pulman, S. G. (2014). Deep learning for answer sentence selection. In *Deep Learning and Representation Learning Workshop: NIPS 2014*. arXiv:1412.1632.

Zafar, H., Napolitano, G., & Lehmann, J. (2018). Formal query generation for question answering over knowledge bases. In *European Semantic Web Conference* (pp. 714–728). Springer.

Zhu, S., Cheng, X., & Su, S. (2020). Knowledge-based question answering by tree-to-sequence learning. *Neurocomputing, 372*, 64–72. ISSN:0925-2312. https://doi.org/10.1016/j.neucom.2019.09.003.

Chapter 2
History and Architecture

Abstract In this chapter, a brief history of QA systems will be introduced. In addition, different components of a QA system will be described in detail. Components of the QA architectures which are related to the main focus of this book will be outlined, and the main QA models in these architectures will be determined. We are going to cover history and architecture for each of the TextQA, KBQA, as well as hybrid systems separately.

2.1 Introduction

In the early stage of developing QA systems, most systems aimed to find answers from knowledge bases or databases by converting natural language questions to SPARQL/SQL queries. Such systems were able to answer questions with high confidence and accuracy. These systems, however, could not answer questions from different domains, since the knowledge bases and databases were limited. In the next stage of this research area, researchers tried to extract answers from text in order to make QA systems domain independent. Although this generation of QA systems overcome the domain dependency problem, they suffer from the accuracy problem. Finding answers from unstructured data such as Web pages requires various techniques including natural processing, information retrieval, information extraction, and text mining which makes the systems difficult and complex. Developing new knowledge bases in the last decades motivated researchers to go back to the first stage of QA systems where they used structured data to answer questions. This backtrack, however, does not lead to the same problem as previous systems. Because the number of knowledge sources is much wider than before and using all existing linked open data provides the possibility of developing open-domain QA systems over knowledge bases. Considering the advantages of both systems, namely, TextQA and KBQA, motivated researchers to go through the next generation of QA systems which benefits from both resources.

In this chapter, we briefly overview the history of QA systems in all generations. The detailed description of main, state-of-the-art QA systems will be described in

S. Momtazi, Z. Abbasiantaeb, *Question Answering over Text and Knowledge Base*,
https://doi.org/10.1007/978-3-031-16552-8_2

Chaps. 5, 6, and 8 for TextQA, KBQA, and real applications of QA including hybrid systems, respectively.

2.2 Closed-Domain Systems with Structured Data

All early-developed QA systems were working on structured data which were limited to specific domains.

BASEBALL (Green et al., 1963) is one of the earliest QA systems. This closed-domain system was developed to answer user questions about dates, locations, and the results of baseball matches by deriving the answers from a database. Some example questions that BASEBALL was able to answer them are as follows:

- *Who did the Red Sox lose to on July 5?*
- *How many games did the Yankees play in July?*
- *On how many days in July did eight teams play?*

In BASEBALL, the input question is analyzed and converted to a canonical form using linguistic knowledge. A query is then generated based on the converted format and is executed over the structured database. BASEBALL was the pioneer in developing QA systems which aim to provide natural language-based front-ends to work with databases.

LUNAR (Woods, 1977) is another early closed-domain QA system. This system was developed to answer natural language questions about the geological analysis of rocks returned by the Apollo moon missions. This system was very effective in its domain; a version of this system demonstrated at a lunar science convention in 1971 was able to answer 90% of questions in its domain posted by people not trained on the system.

Some example questions on LUNAR are as follows:

- *What is the average concentration of aluminum in high alkali rocks?*
- *How many Brescias contain Olivine?*

After BASEBALL and LUNAR, further closed-domain QA systems were developed in the following years. The STUDENT system (Winograd, 1977) was built to answer high school students' questions about algebraic exercises. PHLIQA (Bronnenberg et al., 1980) was developed to answer the user's questions about European computer systems. UC (Unix Consultant) answered questions about the Unix operating system. LILOG was able to answer questions about tourism information in German cities.

The common feature of all these systems is that all of them used a manually created database or knowledge base to find answers to the user's questions. The creators of such systems assumed that informative sources of information are available in structured format, but all users are not able to interact with such data sources; i.e., it is time-consuming and computationally challenging for users to execute their queries over structured databases using a specific query language, such as SQL or SPARQL. Therefore, it is needed to provide an interface where users can interact with data using their own natural language. Such systems help users to find their desired information with no expertise in communicating with advanced structured data sources.

On the one hand, using databases or knowledge bases leads to highly accurate systems. In addition, the systems were very efficient at runtime, since they only need to convert input natural language questions to SQL and SPARQL queries. Considering that these systems are closed-domain and cover limited questions, converting questions to SQL or SPARQL queries is straightforward, and the systems report high accuracy as a result. On the other hand, creating such knowledge sources is done manually which was labor-intensive. Therefore, the existing knowledge sources in the first generation covered very limited domains and could not be expanded to open-domain systems.

2.3 Open-Domain Systems with Unstructured Data

To build more advanced domain-independent systems, researchers tried to expand QA systems to find answers from unstructured data, such as Web documents. They put much effort into building open-domain QA systems, which do not need a database or knowledge base; i.e., the systems should be able to deal with plain texts.

To this aim, the systems have to process the user's question, then retrieve the documents which are related to the question and select the relevant paragraphs/sentences, and finally extract the answer from the relevant paragraphs/sentences. As a result, QA systems require various components based on natural language processing and information retrieval techniques.

By the mid-2000, various QA systems were developed which worked based on Web search. One important example of TextQA systems is OpenEphyra which is available in SourceForge.[1] It provides an open-source framework to build a QA system.

Open-domain QA systems over text became more popular in 1999, when the Text REtrieval Conference (TREC) included a QA track for the first time (Voorhees & Harman, 1999), aiming to encourage researchers to focus on this area of research. The benchmarking of QA was also raised by the Cross-Language Evaluation Forum (CLEF) and NII Test Collections for IR systems (NTCIR) campaigns. A short history of TextQA within TREC is described in Sect. 2.3.1.

[1] http://sourceforge.net/projects/openephyra/.

2.3.1 QA at TREC

Considering that the main requirements to coordinate and compare many research groups working on information retrieval are a standard test set and a large data collection, these requirements motivated people to develop a forum to compare the results of different systems using standard test collections and evaluation metrics. This project in turn led to the TREC conference co-sponsored by the National Institute of Standards and Technology (NIST) and Defense Advanced Research Projects Agency (DARPA) as part of the TIPSTER text program. This conference, which was started in 1992, aims to support research within the information retrieval community and prepare the infrastructure necessary for the large-scale evaluation of text retrieval technologies (Voorhees & Harman, 2007).

The main goals of the conference as mentioned by Voorhees and Harman (2007) are as follows:

- "to encourage research in information retrieval based on large test collections
- to increase communication among industry, academia, and government by creating an open forum for the exchange of research ideas
- to speed the transfer of technology from research laboratories into commercial products by demonstrating substantial improvements in retrieval methodologies on real-world problems
- to increase the availability of appropriate evaluation techniques to be used by industry and academia, including development of new evaluation techniques more applicable to current systems."

The QA track has been run from 1999 to 2007 and has evaluated many QA systems around the world including the Alyssa QA system from Saarland University (Shen et al., 2007), the Ephyra QA system built in Carnegie Mellon University and Karlsruhe University (Schlaefer et al., 2007), the PowerAnswer system from the Lymba Corporation (Moldovan et al., 2007), the FDUQA system developed at Fudan University (Qiu et al., 2007), the CHAUCER system presented by the Language Computer Corporation (LCC) (Hickl et al., 2007), the Pronto system from the University of Rome and University of Sydney (Bos et al., 2007), and the QA systems developed at Concordia University (Razmara et al., 2007), Queens College (Kwok & Dinstl, 2007), and the University of Lethbridge (Chali & Joty, 2007).

After 2007, a new version of the task, named *opinion question answering*, was run in the newly established Text Analysis Conference (TAC) in 2008.

Each year, a data collection and a set of input queries are provided for TREC by NIST. The first data collection used for the QA track contained about 528,000 news articles. This dataset was used for 3 years, and the task was to find complete answer strings to the input questions, with the answer strings limited to a length of 50 or 250 bytes. After that, in 2002, a new data collection, called AQUAINT1, was introduced. The AQUAINT1 corpus[2] contains more than 1,000,000 news articles extracted from 3 sources: the Xinhua News Service (XIN), the New York Times News Service (NYT), and the Associated Press Worldstream News Service (APW). In addition to changing the data collection, the answering policy was also changed. Instead of producing answer strings of a specific length, systems now had to provide a single exact answer. This structure remained for the further years of the QA track at TREC.

Over the 8 years that TREC included the QA, three different question classes were asked during the competition:

- **Factoid questions** ask about a fact and can be answered with a simple short statement, for example, "In what year was Albert Einstein born?"
- **List questions** ask about a list of different instances of a factoid question, for example, "List the names of books written by Herman Hesse."
- **Definition questions** ask for more detailed information and are normally answered by a longer text, such as a full sentence, for example, "What is Lambda calculus?"

In 2008, when the QA track moved to TAC, a new type of questions was posed. This new type, called "opinion questions," focuses on questions that ask about people's opinions, for example, "What do students like about Wikipedia?" and "What organizations are against universal health care?". The motivation for opinion QA comes from the desire to provide tools for individuals, governmental organizations, commercial companies, and political groups who want to automatically track attitudes and feelings in online resources.

2.3.2 Architecture of TextQA

The goal of TextQA is to retrieve a segment of a text or sentence that answers the question. To this aim, usually information retrieval methods are used for extracting the smaller number of documents and reducing the search space. The retrieved documents are the most potential documents to contain the answer sentence. Then, the answer candidates are extracted from these documents, and ranking models are employed for ranking the answer candidates based on the probability of being a correct answer. The ranking model or answer sentence selection model is the most important component of the TextQA, and current researches mostly focus on this task.

Some of the available studies on TextQA focus on the answer sentence selection component; i.e., they receive a set of sentences and rank them based on their relevance to the input question. The top ranked sentences are then selected as

[2] See catalog number LDC2002T31 at http://www.ldc.upenn.edu/Catalog.

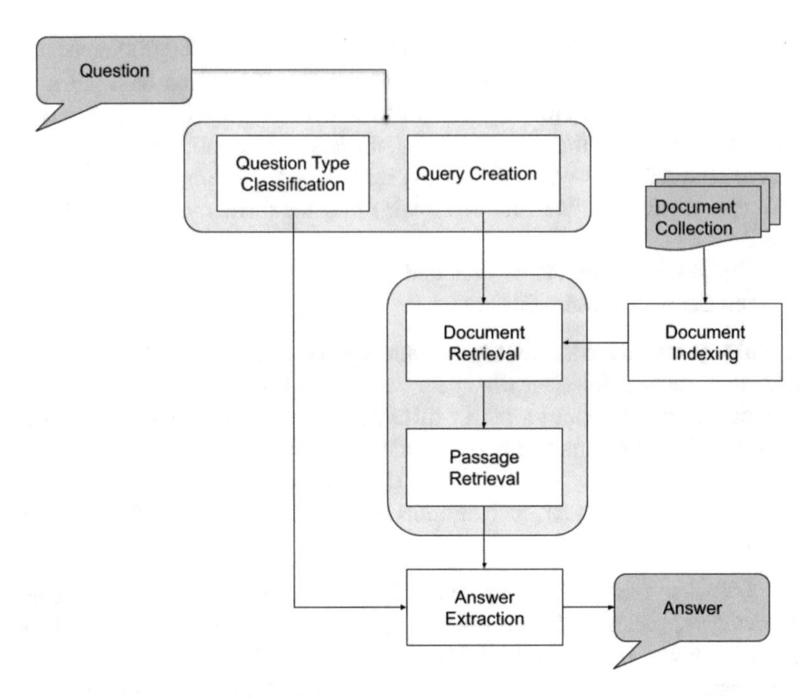

Fig. 2.1 General architecture of TextQA

answers, and they are evaluated by using the datasets that provide answer candidates for each question.

Another group of TextQA systems, however, consists of several components that are used for extracting answer candidates from available documents in the general pipeline of a complete QA model. We will review the general architecture in this chapter and discuss the available models for answer sentence selection in Chap. 5.

General architecture of TextQA is illustrated in Fig. 2.1. This architecture includes three main components including question processing, passage retrieval, and answer processing.

- **Question processing:** The goal of the question processing component is to detect the type of answer and create a query for retrieving the relevant documents. The query is constituted through query formulation, and the type of the answer is determined by question classification.

  1. **Query creation:** In the query formulation process, a list of keywords is extracted from the query sentence and used for querying. The process of forming the query mainly depends on the application of the QA system. When a query is executed over the Web, (1) the stop words are eliminated by search engine, (2) the question words are removed, and (3) the remaining words are selected as keywords. The keywords can be selected from noun phrases by removing the high-frequency and low-content words.

When we want to execute the query over small size corpora, the query expansion techniques are needed, due to the word mismatch problem. In large corpora such as the Web, however, the problem is less pronounced. Therefore, in QA systems, query expansion is required to enhance the probability of retrieving the answer from corpus.

The query reformulation rules are used for rephrasing the question into a declarative answer. For instance, the query "when was the laser invented?" is reformulated to declarative answer "the laser was invented" by applying the rules. Some examples of these rules are as follows:

wh-word did A *verb* B → .. A *verb+ed* B

Where is A → A is located in

2. **Question type classification:** The goal of question classification or answer type recognition task is to determine the type of the entity expected to occur in the answer sentence. For example, when a question asks "Where was X born?", we expect a location in the answer; or when the question asks "Who invented Y?", we realize that the answer must contain the name of a person. Knowing the answer type helps to reduce the size of potential answers. For example, given the question "Where was X born?", we know that the answer must include a location, and as a result, we can ignore sentences without location and stop processing them. In this way, only the sentences with the name of a location are considered as answer candidates. Therefore, not only the required computation but also the noise passed to the next components is decreased.

   The definition questions can be answered according to templates. Consider the definition question "What is a prism?" which its answer can begin with "A prism is ..." and the question "Who is Zhou Enlai?" which requires a biography-specific template including nationality, date of birth, and place of birth.

   In some questions, the answer type is determined by using named entity types, but determining the type of some questions requires using answer type taxonomy or question ontology. The question classification problem can be solved by using hand-built rules or supervised machine learning models. The supervised question classification models are trained over labeled datasets that include the questions and their corresponding answer type. The Webclopedia QA Typology (Hovy et al., 2002) includes 276 hand-written rules for detecting 180 different question types. An example of a rule used for detecting the BIOGRAPHY type is as follows where X must be a Person:

   who {is | was | are | were} X ?

- **Passage retrieval:** The query produced in the previous step is executed, and a sorted list of documents is returned. The documents are not proper for further processes, and they need to be broken into passages. The documents can be broken into passages by using paragraph segmentation algorithms.

  After extracting the passages, the passage retrieval models are executed to rank them according to the probability of including the answer of the given question. At first, the passages that with a high probability do not include the answer are eliminated, and then the passage retrieval is performed over the remaining passages using different approaches. Some features extracted and used for passage retrieval are as follows:

- The count of named entities with the same type as answer type in the passage
- The number of times the question keyword occurs in the passage
- The longest exact match between question keywords and the passage
- The rank of the associated document which included the passage
- The n-gram overlaps
- The proximity score of keywords

By using the Web search for retrieving the relevant documents, the document snippets can be used as passages, and no passage extraction model is required.

- **Answer processing:** The last step of the QA system is to find the answer from passage. Various algorithms and models can be used for selecting the answer from input passage. Some neural or language model-based models can be used for calculating the probability of being a correct answer for each sentence of the input passage. These models extract the sentence containing the correct answer from the input passage. Another example of algorithms used for selecting the answer works by using pre-defined patterns for each answer type. Given the Expected Answer Type (EAT), the pattern extraction methods utilize the specific pre-defined patterns for matching the passage and retrieving the answer.

  Consider the following example with a HUMAN answer type and the given answer sentence. Named entities of the passage are extracted and considered as the answer of the question.

  "Who is the prime minister of India"

  Manmohan Singh, Prime Minister of India, had told left leaders that the deal would not be renegotiated.

  Finding answers to some questions is more complicated than the above example. For instance, the DEFINITION answer types require matching the passage with pre-defined patterns. The patterns can be extracted by humans or be learned by machine.

  The pattern extraction models work by finding the pattern among answer types and a specific aspect of the question sentence. For example, given the answer type "YEAR-OF-BIRTH," the name of the person whose year of birth is asked is the specific aspect of the question required for extracting the pattern. Therefore, the pattern between the answer type and name of the person PERSON-NAME is extracted.

  Different algorithms and models can be used for extracting the patterns associated with each answer type, and the precision of each pattern is calculated. The precision is calculated according to the number of times that the pattern extracted the correct answer. Then, the high-precision patterns are selected.

  The explained models are not sufficient for answering the questions. For example, it is possible that more than one name will be extracted with the same

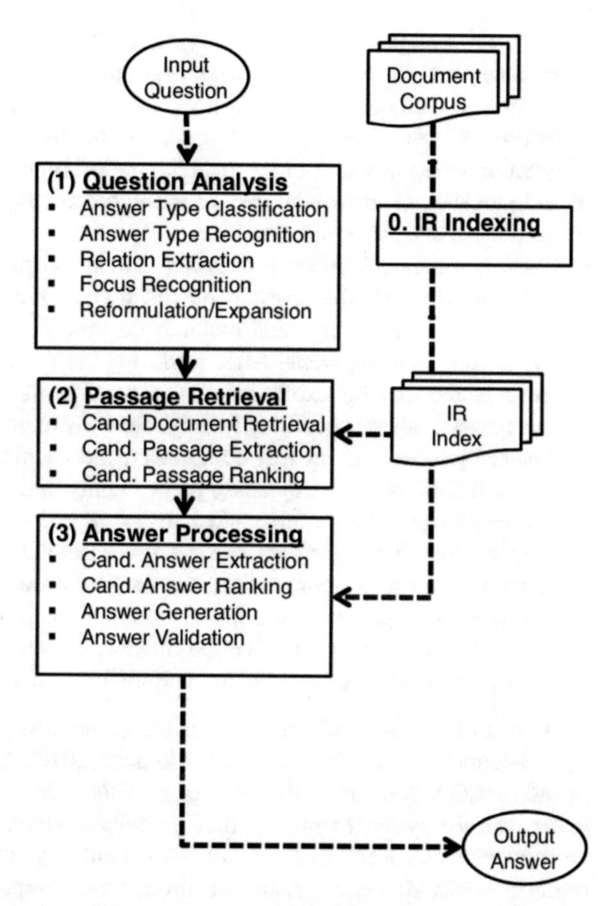

Fig. 2.2 An alternative architecture for TextQA (Dimitrakis et al., 2020) ("©Springer Science+Business Media, LLC, part of Springer Nature, reprinted with permission")

type as the answer type. To overcome the mentioned problem, a combination of both entity detection and pattern extraction models with classifiers is used for ranking the candidate answers. The classifiers use features like existence of the answer type in the candidate answer, the pattern which matches the candidate answer, number of question terms that occurred in the candidate answer, distance between candidate answer terms and query words, novelty of candidate answer sentence, and length of the longest phrase which is common in the question and the candidate answer for ranking the candidate answers.

Another alternative architecture for TextQA is shown in Fig. 2.2. This architecture contains three components including question analysis, passage retrieval, and answer processing. These components are explained briefly in the following.

- **Question analysis:** This component includes classification of question type, detection of question focus, recognition of EAT, extraction of relations, and expansion of the question or reformulating question. EAT identifies the type of the expected entity and is helpful for the list and the factoid questions. There are several approaches for identifying the EAT including rule-based models, machine learning classifiers, deep learning models, and combination of the mentioned methods. Relation extraction means extracting the relation between named entities of the question sentence. In the query formulation task, the question is rephrased in order to add extra information for enhancing the recall and precision of the model. For instance, the synonym words can be extracted from WordNet and used here.

- **Passage retrieval:** After processing the input question and expanding the question sentence, the relevant documents are retrieved to find the relevant information. This component includes document retrieval, passage extraction, and passage ranking steps. After retrieving the relevant documents, the search space is reduced by extracting the relevant passages. Finally, the candidate passages are ranked according to the probability of counting the correct answer.
- **Answer processing:** In this component, the candidate answer sentences are retrieved from passages by using named entity and part-of-speech (POS) tags. The candidate answers are ranked by using a set of features. Rule-based or machine learning classifiers can be used for the task of candidate answer ranking. In the answer generation step, the final answer is generated from several sentences. This step is useful in the case that the correct answer does not exist in one sentence and a combination of several sentences answers the question. Finally, the answer is validated to measure the system's confidence.

As mentioned, TextQA includes various components. Such complexity in TextQA systems decreases their accuracy (Momtazi, 2010). As a result, the performance of the first QA systems at the first stage of their developments was significantly lower than the systems from the first generation which used structured data. The performance, however, increased later by simplifying the pipeline and focusing on retrieving relevant sentences from a limited search space, rather than extracting exact answers from a large collocation of documents.

2.4 Open-Domain Systems with Structured Data

Comparing the performance of the first generation of QA systems, which worked on structured data, with the systems working on unstructured data shows that answers found from knowledge bases are more accurate than those extracted from raw texts. Although they could produce accurate results, they were not used very often after 1970 due to their limited coverage. The main problem of having limited knowledge bases was the manual process of creating such knowledge sources. This problem, however, has been solved by using information extraction techniques, which can automatically extract information from Web documents and create knowledge bases.

Information extraction consists of two main components: (1) a named entity recognizer which aims to find named entities in the text and (2) a relation classifier which detects the type of relation between entities. Having two entities in a text, the classifier finds their relation and inserts them to a knowledge graph, such that the entities are the nodes and the relation is the edge between them.

The START (SynTactic Analysis using Reversible Transformations) system (Katz, 1997) was the first attempt of this kind. This system utilizes a knowledge base to answer the user's questions, but the knowledge base was first created automatically from unstructured Internet data and then used to answer natural language questions. START is the first attempt toward an open-domain QA system, but still with limited capabilities. START was developed by Boris Katz at MIT's Artificial Intelligence Laboratory, and the system is continuously running since 1993 through the Web. The system receives questions in natural language and parses and matches them against its knowledge source. The relevant information is then extracted from the knowledge source to answer the user's question.

The main technique that has been used in START is natural language annotation which provides a bridge between available information sources and users who seek such information. In this model, natural language segments are used as a descriptive form of content that are associated with information segments; an information segment is selected if its annotation in natural language matches the textual content of the input question.

START consists of two main natural language processing modules as follows:

- The understanding module: This module processes text data and creates a knowledge base from the extracted information.
- The generating module: This model uses the appropriate segment of the knowledge base and generates an English sentence as answer.

In early 2000, knowledge bases grew significantly and provided a great opportunity for open-domain KBQA. Wikidata (Vrandečić & Krötzsch, 2014), Freebase (Diefenbach et al., 2018), DBpedia (Lehmann et al., 2015), and YAGO (Suchanek et al., 2008) are examples of these knowledge bases. The existence of such knowledge sources is a great motivation to revisit the first generation of QA systems which used structured data. This time, however, researchers are not limited to a small number of hand-crafted data sources and can use a large number of knowledge bases automatically created from the Web data.

Open-domain QA over structured data has received more attention by the introduction of the QALD[3] (Usbeck et al., 2017) benchmark in 2011. This generation of QA systems became more popular after introducing other benchmarks in the field, including WebQuestions (Berant et al., 2013) and SimpleQuestions (Bordes et al., 2015) which run queries over Freebase.

QALD started in 2011 on two different datasets, namely, DBpedia and MusicBrainz, while focusing on the English language only. The benchmarked continued in the next years by expanding to other languages, including German, Spanish, Italian, French, and Netherlandish. The task also has been expanded in terms of dataset, such that LinkedSpending as well as SIDER, Diseasome, and

[3] http://www.sc.cit-ec.uni-bielefeld.de/qald/.

DrugBank have been added to QALD. Squall2sparql (Ferré, 2013) and CANaLI (Mazzeo & Zaniolo, 2016) are examples of systems participated in the benchmark. Moreover, some systems provide additional options to guide users in formulating their queries; e.g., Sparklis (Ferré, 2017) and Scalewelis (Guyonvarch & Ferré, 2013) proposed a graphical interface for this goal.

2.4.1 Architecture of QA over Knowledge Graph

The KBQA can answer a question if its correct answer exists in the knowledge graph. The information is stored in RDF format in a knowledge graph. Knowledge graphs store entities and their relations by facts. Each fact includes two entities and one relation which connects the head entity to the tail entity. The entities are represented by knowledge graph nodes, and relations are represented by directed edges between entities. The goal of the KBQA is to extract the facts containing the correct answer given the question. A general architecture of this type of QA is depicted in Fig. 2.3. According to Fig. 2.3, KBQA is performed in three main steps including question analysis, query forming, and answer detection that are explained below.

- **Question analysis:** The input question must be converted to a SPARQL query for retrieving the relevant fact from the knowledge graph. To form the SPARQL query which looks for the answer of the question, the head entity and the relation must be captured from the input question. In other words, the input question includes the head entity and the relation, and the answer is the tail entity in

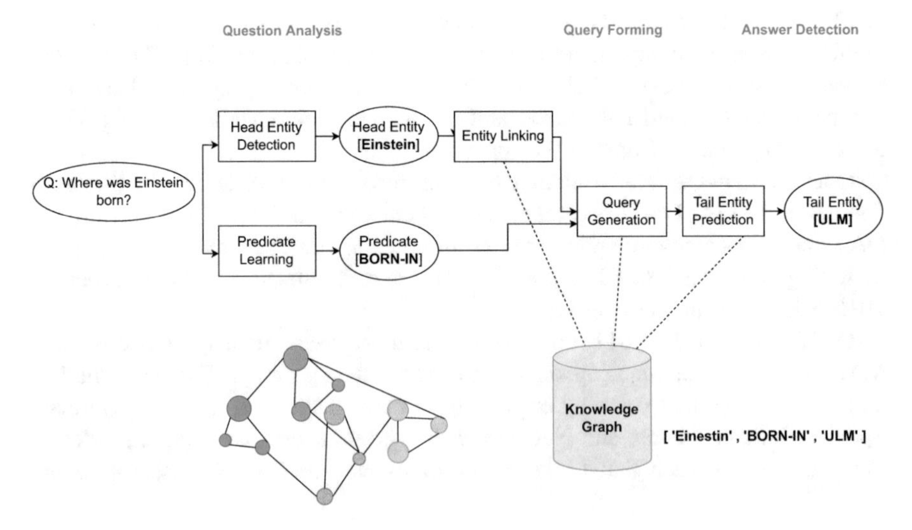

Fig. 2.3 General architecture of KBQA

the fact containing the corresponding head entity and relation. For example, considering the question "Where was Einstein born?", we realize that the head entity is "Einstein" and relation is "BORN-IN." Therefore, the answer is stored in a fact containing the mentioned head entity and relation.

The question analysis component includes two main components, namely, head entity detection and predicate learning. The goal of the head entity detection model is to detect the head entity given the question sentence. The predicate learning model extracts the predicate (or relation) from input questions. Different neural network models and non-neural models are proposed for these two tasks.

- **Query forming:** An entity linking model is utilized to map the detected head entity to an entity (or list of entities) from a knowledge graph. After linking the head entity to the corresponding node(s) in the knowledge graph and detecting the type of relation in the input question, the SPARQL query is generated. SPARQL is a type of query used for querying the graphs.
- **Answer detection:** In this step, the generated query is executed over the knowledge graph to retrieve the relevant fact. The tail entity of the retrieved fact is the answer to the question.

Another general pipeline for KBQA is shown in Fig. 2.4. The pipeline includes one preprocessing step which is executed in offline mode. In this pipeline, the input question is processed and answered through five major steps that are explained in the following.

- **Data preprocessing:** The main goal of this step is to enhance the speed of search over knowledge graphs and increase the accuracy of models in answering the questions. Speed of search is increased by generating index over knowledge base. The vocabulary of entities and relations within the knowledge graph is extended by adding synonyms and abbreviations of the entities and relations. This results in increasing recall in further steps because the entities will be expressed by more names. As the increase in recall is associated with decrease in precision, the trade-off between both sides should be taken into consideration. For example, some names refer to multiple entities, and by adding such names to all relevant entities, the probability of extracting the relevant entity will increase, while more non-relevant entities are extracted.
- **Question analysis:** Analyzing a question includes POS tagging, parsing to capture syntactic structure of question, and named-entity recognition (NER). The output of question analysis is used to determine the question type, focus of the question, and EAT.
- **Data matching:** This step aims at matching the words or phrases of the input question to knowledge graph data. In other words, data matching means finding words or phrases from questions that are equal to an entity in a knowledge graph. The matching is challenging due to the lexical gap and difference between the question meaning and the vocabulary of graph. Different approaches from simple string matching models to more complicated word embedding-based models are proposed for data matching. To match the question sentence with knowledge graph entities, the synonym, hyponym, and hypernym relations are used.

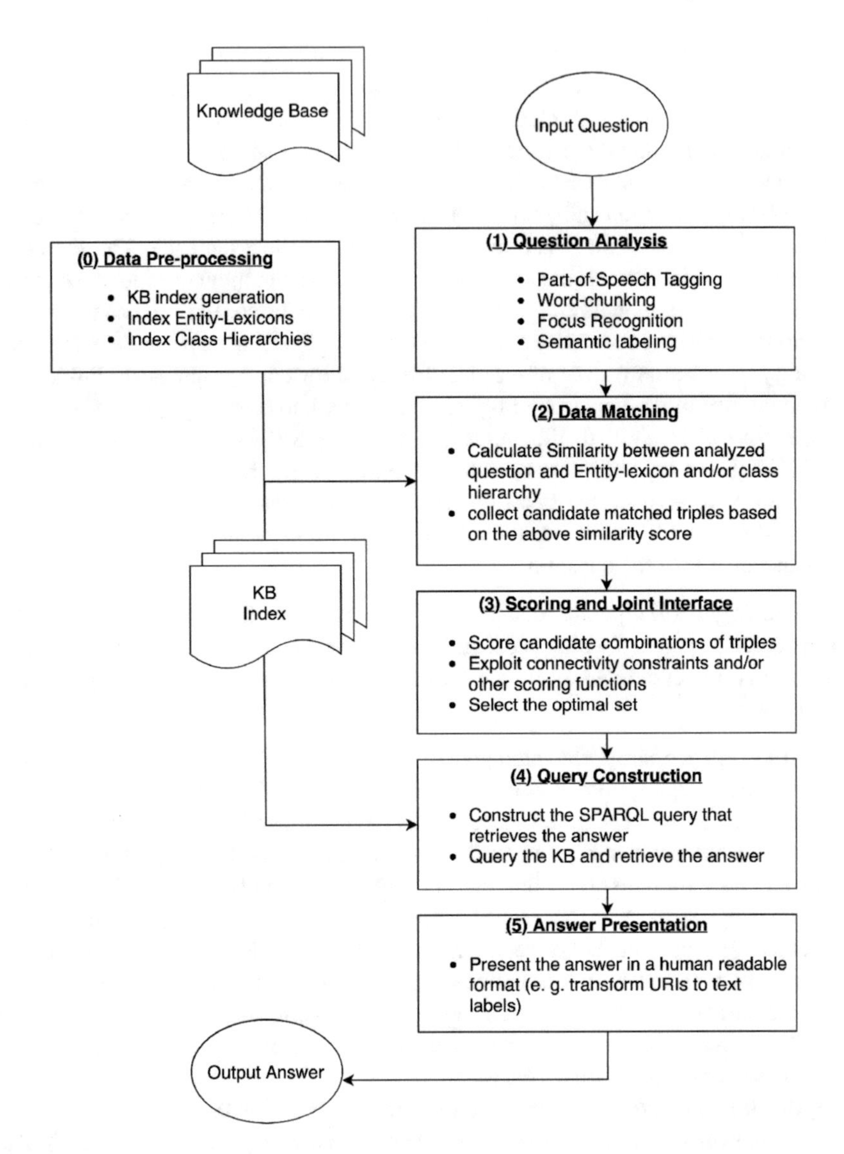

Fig. 2.4 Architecture of QA using knowledge graph (Dimitrakis et al., 2020) ("©Springer Science+Business Media, LLC, part of Springer Nature, reprinted with permission")

The knowledge graphs are equipped with this information, while the question sentence lacks such relations. Language resources or word representation models, such as WordNet (Miller, 1995), PATTY (Nakashole et al., 2012), ReVerb (Fader et al., 2011), Word2Vec (Mikolov et al., 2013), Glove (Pennington et al., 2014), and ELMO (Peters et al., 2018), are used for capturing words' relations.

- **Scoring and joint inference:** A list of candidate components from the knowledge graph is available in this step. Different combinations of these candidate components create a set of candidate answers. The final answer is selected from the list of candidate answers. To this aim, the candidate answers are evaluated and scored, and the answer with the highest score is selected.

  Merging the components has some limitations including (1) lack of connectivity and (2) conflicting components. In the case of conflicting components, one component must be chosen according to the computing score to form the SPARQL query.
- **Query construction:** The corresponding SPARQL query for retrieving the answer of input question is formed. The template-based approaches generate the query by mapping the input question to a set of pre-defined SPARQL query templates. The template-free models generate the SPARQL query according to the syntactic structure of the given question.
- **Answer presentation:** As the retrieved RDF fact from the knowledge graph is not easy to read for users, the RDF fact is converted to a sentence in natural language.

  The factoid questions are answered easily by returning the value of predicate (rdfs:label), while the complete questions are answered by using more processing. Cao et al. (2011) used clustering and summarization techniques for generating the answer from RDF facts.

2.5 Hybrid Systems

Considering the advantages of both types of open-domain QA systems, in recent years, researchers focused on hybrid QA systems, i.e., QA models using both texts and knowledge bases, to benefit from both approaches. HAWK (Usbeck et al., 2015) and YodaQA (Baudi, 2015) are examples of such systems.

Moreover, real application of QA also focused on this type. SIRI[4] as a personal assistant funded by DARPA in 2003 is a hybrid system. The project continued later as an Apple app and integrated in iOS 4 from 2010. Meanwhile, the "Watson" project in IBM followed the same approach, which will be described in more detail in Sect. 8.1. *Alexa*,[5] *OK Google*,[6] *WolframAlpha*,[7] and *Microsoft Cortana*[8] are other big players in the field which benefit from multiple sources, such as free text, Web services, and knowledge bases, to answer user's questions.

---

[4] http://www.apple.com/ios/siri/.

[5] https://developer.amazon.com/alexa.

[6] https://support.google.com/websearch/answer/2940021?hl=en.

[7] http://www.wolframalpha.com.

[8] http://windows.microsoft.com/en-us/windows-10/getstarted-what-is-cortana.

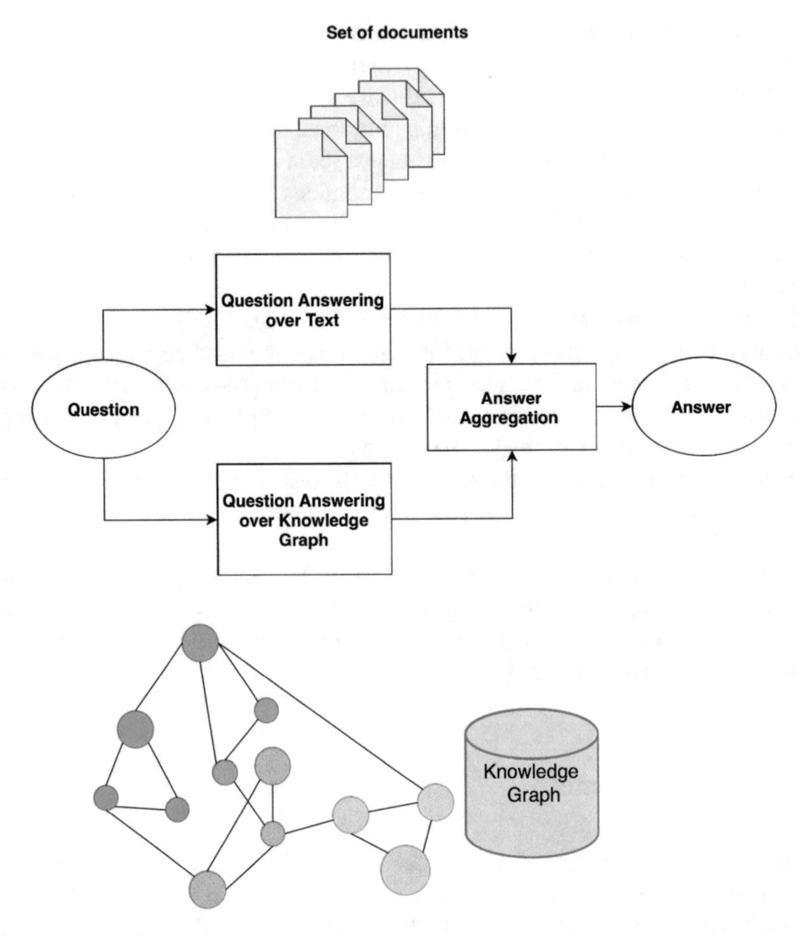

Fig. 2.5 A hybrid architecture for the QA task by combining both the TextQA and KBQA approaches

2.5.1 Architecture of Hybrid QA

A hybrid architecture uses several data sources for answering the input question. This architecture is useful when the available data source is not enough for extracting the answer. By combining various data sources, the chance of extracting the correct answer increases. In the hybrid architecture, which is depicted in Fig. 2.5, both the text documents and the knowledge graphs are utilized for extracting the correct answer. Some of the approaches utilize the text documents to improve the vocabulary of knowledge graph (Savenkov & Agichtein, 2016; Xu et al., 2016a,b).

Some of the other approaches capture data from various data sources and transform it into a unique representation like universal schema to use it in the knowledge graph (Das et al., 2017). The second approach is suitable for end-to-end neural architectures as it needs a single data representation. The second approach does not deal with the challenge of adding unstructured text to knowledge graph RDFs.

2.6 Summary

We devoted this chapter to provide a history of QA systems along with their architecture. The closed-domain QA systems which use structured data are studied in Sect. 2.2. The open-domain systems with unstructured and structured data are introduced in Sects. 2.3 and 2.4. Finally, the hybrid QA systems are explained in Sect. 2.5.

References

Baudi, P. (2015). Yodaqa: A modular question answering system pipeline. In *International Student Conference on Electrical Engineering*.

Berant, J., Chou, A., Frostig, R., & Liang, P. (2013). Semantic parsing on Freebase from question-answer pairs. In *Proceedings of the 2013 Conference on Empirical Methods in Natural Language Processing*, Seattle, Washington, USA (pp. 1533–1544). Association for Computational Linguistics. https://www.aclweb.org/anthology/D13-1160.

Bordes, A., Usunier, N., Chopra, S., & Weston, J. (2015). Large-scale simple question answering with memory networks. Preprint. arXiv:1506.02075.

Bos, J., Guzzetti, E., & Curran, J. (2007). The pronto QA system at TREC 2007: Harvesting hyponyms, using nominalisation patterns, and computing answer cardinality. In *Proceedings of the Text REtreival Conference (TREC)*.

Bronnenberg, W., Bunt, H., Landsbergen, S., Scha, R., Schoenmakers, W., & van Utteren, E. (1980). The question answering system PHLIQA1. In L. Bolc (Ed.), *Natural language question answering systems*. Macmillan.

Cao, Y., Liu, F., Simpson, P., Antieau, L., Bennett, A., Cimino, J. J., Ely, J., & Yu, H. (2011). Askhermes: An online question answering system for complex clinical questions. *Journal of Biomedical Informatics, 44*(2), 277–288.

Chali, Y., & Joty, S. (2007). University of Lethbridge's participation in TREC 2007 QA track. In *Proceedings of the Text REtreival Conference (TREC)*.

Das, R., Zaheer, M., Reddy, S., & McCallum, A. (2017). Question answering on knowledge bases and text using universal schema and memory networks. Preprint. arXiv:1704.08384.

Diefenbach, D., Lopez, V., Singh, K., & Maret, P. (2018). Core techniques of question answering systems over knowledge bases: a survey. *Knowledge and Information Systems, 55*(3), 529–569. ISSN:0219-3116.

Dimitrakis, E., Sgontzos, K., & Tzitzikas, Y. (2020). A survey on question answering systems over linked data and documents. *Journal of Intelligent Information Systems, 55*(2), 233–259.

Fader, A., Soderland, S., & Etzioni, O. (2011). Identifying relations for open information extraction. In *Proceedings of the 2011 Conference on Empirical Methods in Natural Language Processing* (pp. 1535–1545).

Ferré, S. (2013). squall2sparql: a Translator from Controlled English to Full SPARQL 1.1. In E. Cabrio, P. Cimiano, V. Lopez, A.-C.N. Ngomo, C. Unger, & S. Walter (Eds.), *Work.*

multilingual question answering over linked data (QALD-3), Valencia, Spain. https://hal.inria.fr/hal-00943522. See Online Working Notes at http://www.clef2013.org/.
Ferré, S. (2017). Sparklis: An expressive query builder for SPARQL endpoints with guidance in natural language. *Semantic Web, 8*, 405–418.
Green, B., Wolf, A., Chomsky, C., & Laughery, K. (1963). Baseball: Aan automatic question answerer. In E. Figenbaum, & J. Fledman (Eds.), *Computers and thoughts*. McGraw-Hill.
Guyonvarch, J., & Ferré, S. (2013). Scalewelis: A scalable query-based faceted search system on top of SPARQL endpoints. In P. Forner, R. Navigli, D. Tufis, & N. Ferro (Eds.), *CEUR workshop at CLEF*, vol. 1179 of *CEUR workshop proceedings* (pp. 23–26). CEUR-WS.org.
Hickl, A., Roberts, K., Rink, B., Bensley, J., Jungen, T., Shi, Y., & Williams, J. (2007). Question answering with LCC's CHAUCER-2 at TREC 2007. In *Proceedings of the Text REtreival Conference (TREC)*.
Hovy, E., Hermjakob, U., & Ravichandran, D. (2002). A question/answer typology with surface text patterns. In *Proceedings of the Second International Conference on Human Language Technology Research, HLT '02*, San Francisco, CA, USA (pp. 247–251). Morgan Kaufmann Publishers Inc.
Katz, B. (1997). Annotating the world wide web using natural language. In *Proceedings of the International Conference on Computer Assisted Information Searching on the Internet (RIAO)*.
Kwok, K., & Dinstl, N. (2007). Testing an entity ranking function for English factoid QA. In *Proceedings of the Text REtreival Conference (TREC)*.
Lehmann, J., Isele, R., Jakob, M., Jentzsch, A., Kontokostas, D., Mendes, P. N., Hellmann, S., Morsey, M., Van Kleef, P., Auer, S., et al. (2015). Dbpedia–a large-scale, multilingual knowledge base extracted from wikipedia. *Semantic Web, 6*(2), 167–195.
Mazzeo, G., & Zaniolo, C. (2016). Answering controlled natural language questions on RDF knowledge bases. In *Extending Database Technology (EDBT)*.
Mikolov, T., Sutskever, I., Chen, K., Corrado, G., & Dean, J. (2013). Distributed representations of words and phrases and their compositionality. In *Proceedings of the 26th International Conference on Neural Information Processing Systems - Volume 2, NIPS'13*, USA (pp. 3111–3119). Curran Associates Inc.
Miller, G. A. (1995). Wordnet: A lexical database for english. *Communications of the ACM, 38*(11), 39–41.
Moldovan, D., Clark, C., & Bowden, M. (2007). Lymba's PowerAnswer 4 in TREC 2007. In *Proceedings of the Text REtreival Conference (TREC)*.
Momtazi, S. (2010). *Advanced Language Modeling for Sentence Retrieval and Classification in Question Answering Systems*. PhD thesis, Saarland University.
Nakashole, N., Weikum, G., & Suchanek, F. (2012). Patty: A taxonomy of relational patterns with semantic types. In *Proceedings of the 2012 Joint Conference on Empirical Methods in Natural Language Processing and Computational Natural Language Learning* (pp. 1135–1145).
Pennington, J., Socher, R., & Manning, C. (2014). Glove: Global vectors for word representation. In *Proceedings of the 2014 Conference on Empirical Methods in Natural Language Processing (EMNLP)*, Doha, Qatar (pp. 1532–1543). Association for Computational Linguistics. https://doi.org/10.3115/v1/D14-1162.
Peters, M., Neumann, M., Iyyer, M., Gardner, M., Clark, C., Lee, K., & Zettlemoyer, L. (2018). Deep contextualized word representations. In *Proceedings of the 2018 Conference of the North American Chapter of the Association for Computational Linguistics: Human Language Technologies, Volume 1 (Long Papers)*, New Orleans, Louisiana (pp. 2227–2237). Association for Computational Linguistics. https://doi.org/10.18653/v1/N18-1202.
Qiu, X., Li, B., Shen, C., Wu, L., Huang, X., & Zhou, Y. (2007). FDUQA on TREC 2007 QA track. In *Proceedings of the Text REtreival Conference (TREC)*.
Razmara, M., Fee, A., & Kosseim, L. (2007). Concordia University at the TREC 2007 QA track. In *Proceedings of the Text REtreival Conference (TREC)*.
Savenkov, D., & Agichtein, E. (2016). When a knowledge base is not enough: Question answering over knowledge bases with external text data. In *Proceedings of the 39th International ACM SIGIR Conference on Research and Development in Information Retrieval, SIGIR '16*, New

York, NY, USA (pp. 235–244). Association for Computing Machinery. ISBN:978-1-4503-4069-4. https://doi.org/10.1145/2911451.2911536.

Schlaefer, N., Ko, J., Betteridge, J., Pathak, M., Nyberg, E., & Sautter, G. (2007). Semantic extensions of the ephyra QA system for TREC 2007. In *Proceedings of the Text REtreival Conference (TREC)*.

Shen, D., Wiegand, M., Merkel, A., Kaszalski, S., Hunsicker, S., Leidner, J., & Klakow, D. (2007). The Alyssa system at TREC QA 2007: Do we need Blog06? In *Proceedings of the Text REtreival Conference (TREC)*.

Suchanek, F. M., Kasneci, G., & Weikum, G. (2008). Yago: A large ontology from wikipedia and wordnet. *Journal of Web Semantics, 6*(3), 203–217.

Usbeck, R., Ngomo, A.-C. N., Bühmann, L., & Unger, C. (2015). Hawk – hybrid question answering using linked data. In F. Gandon, M. Sabou, H. Sack, C. d'Amato, P. Cudré-Mauroux, & A. Zimmermann (Eds.), *The semantic web. Latest advances and new domains* (pp. 353–368). Cham: Springer International Publishing.

Usbeck, R., Ngomo, A.-C. N., Haarmann, B., Krithara, A., Röder, M., & Napolitano, G. (2017). 7th open challenge on question answering over linked data (qald-7). In *Semantic web evaluation challenge* (pp. 59–69). Springer.

Voorhees, E., & Harman, D. (1999). Overview of the wigth text REtrieval conference (TREC8). In *Proceedings of the Text REtrieval Conference (TREC8)*.

Voorhees, E., & Harman, D. (2007). *TREC: Experiment and evaluation in information retrieval*. The MIT Press.

Vrandečić, D., & Krötzsch, M. (2014). Wikidata: a free collaborative knowledgebase. *Communications of the ACM, 57*(10), 78–85.

Winograd, T. (1977). Five lectures on artificial intelligence. In A. Zampolli (Ed.), *Linguistic structures processing* (pp. 399–520). North-Holland.

Woods, W. A. (1977). Lunar rocks in natural english: Explorations in natural language question answering. In A. Zampolli (Ed.), *Linguistic structures processing* (pp. 521–569). North-Holland.

Xu, K., Feng, Y., Huang, S., & Zhao, D. (2016a). Hybrid question answering over knowledge base and free text. In *Proceedings of COLING 2016, the 26th International Conference on Computational Linguistics: Technical Papers*, Osaka, Japan (pp. 2397–2407). The COLING 2016 Organizing Committee. https://www.aclweb.org/anthology/C16-1226.

Xu, K., Reddy, S., Feng, Y., Huang, S., & Zhao, D. (2016b). Question answering on Freebase via relation extraction and textual evidence. In *Proceedings of the 54th Annual Meeting of the Association for Computational Linguistics (Volume 1: Long Papers)*, Berlin, Germany (pp. 2326–2336). Association for Computational Linguistics. https://doi.org/10.18653/v1/P16-1220. https://www.aclweb.org/anthology/P16-1220.

Chapter 3
Question Answering Evaluation

Abstract This chapter is devoted to datasets and evaluation metrics for TextQA and KBQA. Some information such as statistical characteristics, data collection, structure, and source of best-known TextQA datasets will be discussed in this chapter. The most popular knowledge bases, as well as question sets which have been widely used as a benchmark in many recent works on KBQA, will be introduced and described. The structure of storing and retrieving information from knowledge bases and statistical characteristics of them will be covered in this chapter. Moreover, different evaluation metrics used for evaluating both types of QA models are explained.

3.1 Evaluation of TextQA

3.1.1 Datasets

In the general structure of datasets for TextQA or the answer selection task, the questions are linked to a set of candidate answers where each candidate answer is labeled as true or false. The available datasets can be categorized into closed-domain and open-domain datasets. Closed-domain datasets include questions about a specific domain, while open-domain datasets contain questions with different subjects and have no restriction on the topic of questions. Among the famous answer selection datasets, we can name TREC-QA, WikiQA, Yahoo!, and MovieQA. A detailed description of the mentioned datasets is provided in the following.

3.1.1.1 TREC-QA

The TREC-QA dataset is collected from the TREC 8–13 tracks. The questions of the TREC 8–12 tracks are used for creating the training set, and TREC 13 is used for creating the validation and the test set. The candidate answers are extracted from the corresponding document pool of questions. The sentences that have at least one common non-stop word with the question sentence or have the pattern

S. Momtazi, Z. Abbasiantaeb, *Question Answering over Text and Knowledge Base*,
https://doi.org/10.1007/978-3-031-16552-8_3

Table 3.1 Statistics of the TREC-QA dataset (Abbasiantaeb & Momtazi, 2021)

	Train-all	Train	Validation (clean)	Test (clean)
# Questions	1229	94	82 (65)	100 (68)
# QA pairs	53,417	4718	1148 (1117)	1517 (1442)
% correct	12.00%	7.40%	19.30%	18.70%
#Answers/Q	43.46	50.19	14.00	15.17
Judgement	Automatic	Manual	Manual	Manual

of expected correct answer are extracted. Questions of validation and test sets and first 100 questions of the training set are judged manually. Other questions from the training set are automatically judged by using regular expressions and matching the pattern of questions with regular expressions.

TREC includes two training sets, namely, TRAIN and TRAIN-ALL. The TRAIN version only includes the manually judged questions from the training set, while the TRAIN-ALL version includes all of the training set questions. A detailed statistics of TREC-QA dataset is shown in Table 3.1. In the clean version of TREC-QA, the questions with no correct or incorrect answer are removed from the dataset.

3.1.1.2 WikiQA

WikiQA (Yang et al., 2015) is collected from Bing query logs. The question-like queries are selected by using simple rules including selecting the queries that start with "wh" question words or have a question mark at the end. Then, the selected queries are filtered to ensure just the questions remain. For example, during the filtering step, the query "How I met your mother" is eliminated as it is the name of a TV series.

To ensure the quality of the dataset, questions that have the following conditions are selected: 1) the question is searched by at least five users, and 2) users have clicked on the wikipedia pages in the returned results of the question. The summary section of the corresponding Wikipedia page is used for retrieving candidate answers. The questions are labeled with crowdsourcing.

For annotating the candidate sentences, at the first, the question sentence with the summary section of the corresponding Wikipedia page is shown to reviewers. If the question is answered in the summary section, they are asked to mark each sentence of the summary section as correct answer or false. And if the summary section cannot answer the question, all of the sentences are labeled as false. Three reviewers are asked to judge the questions, and their judgements are compared. In the case of disagreement between reviewers' decisions, other reviewers are asked to judge the question, and finally, the questions are tagged according to the majority of votes of reviewers.

More statistics of the WikiQA dataset is presented in Table 3.2. In the clean version of the dataset, the questions without any correct answer are eliminated.

Table 3.2 Statistics of the WikiQA dataset (Abbasiantaeb & Momtazi, 2021)

	Train	Validation	Test	Total
# of questions (raw)	2118	296	633	3047
# of questions (clean)	873	126	243	1242
# of sentences	20,360	2733	6165	29,258
# of answers	1040	140	293	1473
Average length of questions	7.16	7.23	7.26	7.18
Average length of sentences	25.29	24.59	24.95	25.15
# of questions w/o answers	1245	170	390	1805

Table 3.3 An example of a question with negative and positive answers for Yahoo! QA

S_X (question)	How to get rid of **memory stick error** of my Sony Cyber-shot?
S_Y^+ (positive answer)	You might want to try to format the **memory stick** but what is the **error** message you are receiving
S_Y^- (negative answer)	Never heard of stack underflow **error**, overflow yes, overflow is due to running out of virtual **memory**

3.1.1.3 Yahoo!

Yahoo! dataset (Wan et al., 2016a,b) is collected from a community QA system named Yahoo! Answers. In Yahoo! Answers, users submit their questions, and other users provide answers to the questions. Then, the user who issued the question selects the best answer to the posted question. 142,627 QA pairs are available at Yahoo! Answerer. The available QA pairs are filtered by restricting the length of the questions and answers in QA pairs between 5 and 50 words. After applying the filter, 60,564 QA pairs remain where all of the questions have correct answers and QA pairs are positive samples. For gathering the negative samples, negative sampling is applied. The best answer of each question is used for querying the whole answer set, and the first 1000 returned answers are selected. Four random answers are selected from the retrieved answers to form the negative samples. At the end, the dataset is divided to train, validation, and test splits. An example of Yahoo! QA dataset is represented in Table 3.3.

3.1.1.4 MovieQA

MovieQA dataset (Tapaswi et al., 2016) is gathered from diverse sources of information including subtitles, plot synopsis, scripts, video clips, and DVS. Plot synopses are summary of movies provided by people who watched them. They depict the events of the movie and interaction of characters in 1 to 20 paragraphs. Video clips are not enough for comprehending the events happening in movies, and they will be completed by subtitles. DVS is a service which explains the movie by adding auxiliary information from visual scenes of the movie to blind people.

Scripts are written by screenwriters and used by characters to build movies. Scripts include more information than dialogs.

The MovieQA dataset includes two modes: (1) plots and subtitles and (2) video clips. For collecting the dates, the plot synopses are used as context of questions. The data collection is performed in two steps described in the following.

- **Question and correct answer:** Annotators can choose a movie from a list of available movies. Then, the plot synopsis of the selected movie is shown to the annotator paragraph by paragraph. The annotator is free to provide any number of questions from the given plot. The annotator is asked to select the correct answer for the question and mark the sentences required for framing the question and answering it. Besides that, the annotators are required to select part of the movie which is about the question and includes its answer.
- **Multiple answer choices:** In this step, the annotators are shown a paragraph and a question and asked to answer the question correctly and provide four incorrect answers to the question. They are allowed to correct the questions or re-state them.
- **Timestamp to video:** In this step, video clips corresponding to each question/answer pair are extracted. The in-house annotators are asked to watch the movies and link each sentence of the plot to a snippet of the movie by selecting the beginning and end in the movie. As each question and answer pair is associated with one or several sentences of plot, the QA pairs are linked to movie plots.

The MovieQA dataset contains 14,944 questions gathered from 408 movies. The questions vary from simple types like "who," "what," and "whom" to more complex questions like "how" and "why." Each question has one correct answer and four incorrect answers in this dataset. More statistics about this dataset is provided in Table 3.4.

3.1.1.5 InsuranceQA

InsuranceQA (Feng et al., 2015) is a closed-domain QA dataset in insurance domain which is collected from the Internet. Each question is linked to an answer pool that contains 500 answers. The whole dataset includes 24,981 distinct answers that are

Table 3.4 Statistics of the MovieQA dataset

	Train	Validation	Test	Total
Movies with plots and subtitles				
# of movies	269	56	83	408
# of QA	9848	1958	3138	14,944
Q # words	9.3	9.3	9.5	9.3 ± 3.5
CA. # words	5.7	5.4	5.4	5.6 ± 4.1
WA. # words	5.2	5.0	5.1	5.1 ± 3.9

Table 3.5 Statistics of the InsuranceQA dataset (Feng et al., 2015)

	Train	Dev	Test1	Test2
Questions	12,887	1000	1800	1800
Answers	18,540	1454	2616	2593
Question word count	92,095	7158	12,893	12,905

used for creating the answer pool of each question. The true answer is added to the answer pool, and other answers are randomly selected from the whole answer set to form the answer pool with size 500. The dataset is divided into train, development, test1, and test2 sets. A detailed statistics of the InsuranceQA dataset is provided in Table 3.5.

3.1.2 Metrics

The goal of the TextQA is to retrieve a sorted list of candidate answers for the input question according to their matching (relevance) score. Hence, the task of TextQA can be seen as an information retrieval problem and can be evaluated using mean reciprocal rank (MRR), P@k, and mean average precision (MAP) metrics. A detailed explanation of these metrics is provided in the following:

- **MRR:** MRR is the mean of reciprocal rank (RR) for each question. The RR of each question is defined as the inverse of the rank of the first retrieved correct answer. As a result, RR emphasizes on retrieving the true answer in the first rank and is not sensitive to the correctness of the retrieved answers in other ranks. RR and MRR are calculated using the following equations:

$$RR(q) = \begin{cases} \frac{1}{\text{the rank of the first correct answer}} & \text{(if a correct answer exists)} \\ 0 & \text{(if no correct answer)} \end{cases} \tag{3.1}$$

$$\text{MRR} = \frac{1}{|Q|} \sum_{q=1}^{|Q|} RR(q) \tag{3.2}$$

where $|Q|$ is the length of the question set.

Considering the example provided in Fig. 3.1, MRR of the input question for model A is 1, because the rank of the first correct answer is 1 and MRR of the input question according to the model B is 0.5 as the first correct answer is in the second place.

- **P@k:** Precision at k-th rank (p@k) is calculated by dividing the count of the true answers among first k retrieved answers to k as follows:

$$\text{Precision@k} = \frac{\text{number of correct answers among first } k \text{results}}{k} \tag{3.3}$$

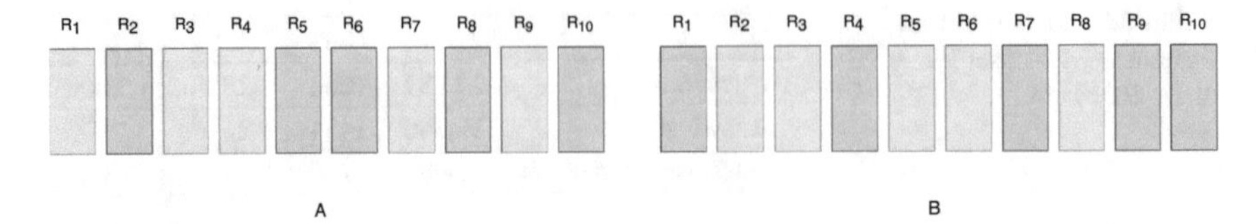

Fig. 3.1 Result of evaluation of a question with five correct and five incorrect answers by two models (A and B). The answers are sorted according to the calculated matching score in descending order. Correct answers are indicated by green and incorrect ones are shown by red color

P@k emphasizes on ranking all of the correct answers higher than the incorrect answers. In an ideal system, having a total number of N true answers among M answers, the value of p@k must be one for $k < N + 1$. Calculating p@5 is a common metric in retrieval-based systems. Considering the example provided in Fig. 3.1, p@5 for both models A and B is 0.6 since three out of five retrieved answers are correct.

- **MAP:** MAP is the mean of the average precision (AP) for each question. Despite the RR, AP measures the ability of the system in retrieving all of the true answers, whereas RR measures the ability of the system to provide one true answer. AP is average of precision at each rank where a correct answer is retrieved. MAP is calculated as follows:

$$\text{AP(q)} = \frac{1}{k} * \sum_{k=1}^{K} \text{P@k} \tag{3.4}$$

$$\text{MAP} = \frac{1}{|Q|} \sum_{q=1}^{|Q|} AP(q) \tag{3.5}$$

where $|Q|$ is the length of the question set and k is the number of the correct answers for each question.

In Fig. 3.2, p@k is calculated for each rank when a correct answer is retrieved. By calculating the p@k values, the average precision of the input question according to evaluation models A and B is as follows:

$$\text{AP}_{\text{model A}} = \frac{\frac{1}{1} + \frac{2}{3} + \frac{3}{4} + \frac{4}{7} + \frac{5}{9}}{5} = 0.70 \tag{3.6}$$

$$\text{AP}_{\text{model B}} = \frac{\frac{1}{2} + \frac{2}{3} + \frac{3}{5} + \frac{4}{6} + \frac{5}{8}}{5} = 0.61 \tag{3.7}$$

- **Accuracy:** This metric is used for evaluating the questions with just one correct answer. It is widely used in KBQA where each question is answered with a fact from a knowledge graph and the object of the fact is the answer of the question.

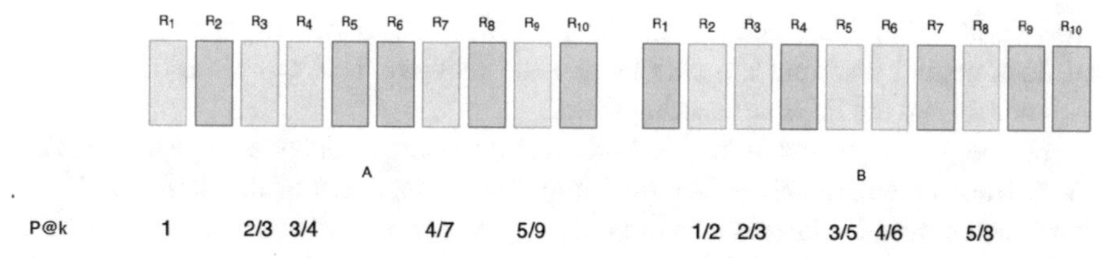

Fig. 3.2 Result $p@k$ at each rank when a correct answer is retrieved for input question with five correct and five incorrect answers by two A and B models. The answers are sorted according to the calculated matching score in descending order. Correct answers are indicated by green and incorrect ones are shown by red color

The metric is also used in TextQA. Accuracy calculates the ability of the model in answering the questions correctly. To be more specific, it indicates how often or with what probability the model answers the given question correctly. Accuracy is portion of the correctly answered question to total number of the questions and is calculated as follows:

$$\text{Accuracy} = \frac{\text{number of correct answered questions}}{N} \tag{3.8}$$

where N is the number of questions.

For example, accuracy of a model which answers 52 questions among 70 input questions correctly is $\frac{52}{70} = 0.74$

3.2 Evaluation of KBQA

KBQA requires a knowledge graph for retrieving the answer of the input questions. For the evaluation of the proposed models, some well-known datasets including WebQuestions, SimpleQuestions, LC-QuAD, and QALD are proposed, and the questions from the mentioned datasets are answered by using a knowledge graph. As a result, we introduce the well-known knowledge graphs used in the task besides the proposed datasets for evaluating the KBQA in the following sections. In addition, the metrics used for evaluation are explained.

3.2.1 Knowledge Graphs

The goal of the Semantic Web is to represent information with a semantic-aware structure on the Web and query them. Also, the Semantic Web is defined as establishing the information with relations visually Freedman and Reynolds (1980). Guns (2013) proposed the idea of representing the semantic with typed links in the World Wide Web. Berners-Lee et al. (2001) suggested expanding the Web of documents by adding the entities and the relations between them besides the documents and the links connecting them.

By the growth of the Semantic Web, many structured datasets appeared on the Web. Knowledge graphs are the most important component of the Semantic Web. Knowledge base includes a specific ontology, which describes a series of classes, and a set of samples from these classes. A knowledge graph is a knowledge base which includes a huge amount of facts describing the entities. Knowledge graphs are large-scale datasets which store information in the format of facts.

The Resource Description Framework (RDF) is a type of constructing semantic graph model used in knowledge graphs, and SPARQL is a language for querying the knowledge graphs with RDF format.

Facts can be stored in triple format as (head_entity, predicate, tail_entity) where head_entity and tail_entity are entities and predicate is relation. Nodes in the knowledge graph are entities, and the directed edges connecting the nodes are relations.

Well-known knowledge graphs used for answering the question in KBQA include Wikidata (Vrandečić & Krötzsch, 2014), Freebase (Diefenbach et al., 2018), DBpedia (Lehmann et al., 2015), and YAGO (Suchanek et al., 2008). Knowledge graphs use entities and predicates for storing the facts. A brief explanation of the well-known knowledge graphs is provided in the following sections.

3.2.1.1 Freebase

Freebase (Diefenbach et al., 2018) is created collaboratively which guarantees the reliability of its facts. Freebase is gathered from structured data of Wikipedia, NNDB, FMD, and MusicBrainz. Freebase has its own API for accessing the data. Entities are recognized by their identifiers (ID) and can be mentioned with other names that are called aliases. Each entity is linked to a set of names called aliases. Vocabulary of relationships and entity types are closed. Besides ID, entities are recognized by URIs. The GraphD graph dataset is used by Freebase which enables it to keep metadata for facts. The architecture of Freebase only permits attaching facts, and the eliminated facts are specified by marks. Users have the freedom to edit Freebase data directly. Freebase includes non-binary relations that are stored by using mediator nodes. An example of a non-binary relation is represented in Fig. 3.3. As is shown in this figure, this feature enables storing a fact like “Portugal is a member of the European Union since 1986.” via a mediator node. Despite DBpedia, Freebase stores the inverse of relationships.

Freebase has two main subsets named FB2M and FB5M that are widely used in QA. FB2M and FB5M include 2M and 5M entities and 5k and 7.5k relations, respectively. Detailed statistics of these two subsets are provided in Table 3.6.

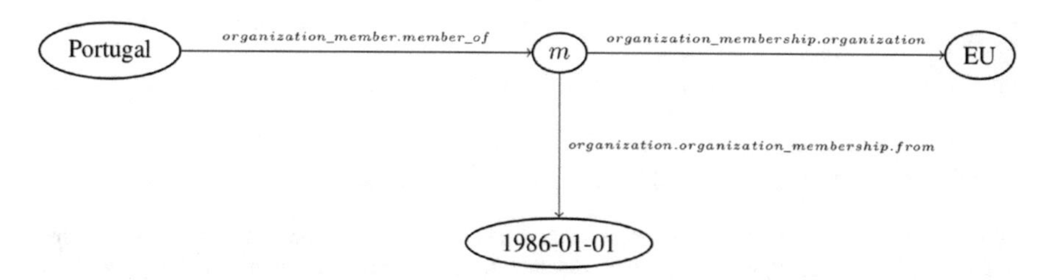

Fig. 3.3 An example of storing non-binary relations in Freebase via a mediator node (Diefenbach et al., 2018) ("©Springer-Verlag London Ltd., part of Springer Nature, reprinted with permission")

Table 3.6 Statistics of two subsets of Freebase knowledge base, namely, FB2M and FB5M (Diefenbach et al., 2018)

	FB2M	FB5M
Entities	2,150,604	4,904,397
Relationships	6701	7523
Atomic facts	14,180,937	22,441,880
Facts grouped	10,843,106	12,010,500

The key features of Freebase are as follows:

- Freebase is dynamic and users have privilege to update it,
- Labels of entities and properties are provided in several languages,
- Each fact has a confidence score which is stored in knowledge graph,
- RDF format is used for dataset,
- Data is accessible via files and HTTP lookup,
- It is part of the Linked Open Data (LOD) cloud which means the principles of Linked Data are followed,
- Textual descriptions of entities and relations are provided,
- Descriptions of classes are provided,
- It has a fixed schema which users can extend it,
- It has its own vocabulary.

3.2.1.2 Wikidata

Wikidata (Vrandečić & Krötzsch, 2014) is another collaborative knowledge graph started in October 2012. Wikidata is gathered and modified collaboratively by users, and its schema is extended by their agreement. The origin of the facts stored in Wikidata is stored and accessible. Wikidata retrieves facts from other sources and keeps their references.

Wikidata stores data in the form of property-value pairs. Each item has a set of property-value pairs where property is an object with its own Wikidata page. For example, the item "Rome" has a property-value pair ("population," 2,777,979). The property "population" has its own Wikidata page which contains its datatype. In addition, each property has labels, descriptions, and aliases. Wikidata stores

Fig. 3.4 An example of storing a complex statement in Wikidata . The main property-value pair is ("base salary," 36,000,000 euro) which is stored by using two other qualifiers ("employer," FC Barcelona) and ("point in time," 2016)

more complex statements which cannot be represented by one property-value pair. For example, the statement "population of Rome was 2,651,040 as of 2010 based on estimations" includes several property-value pairs, namely, ("population," 2,651,040), ("as of," 2010), and ("method," estimation). The two last property-value pairs do not directly refer to "Rome" item because they are required in describing the first property-value pair which is called claim. These property-value pairs are subordinate to the claim property-value pair and are called qualifiers. Example of storing a complex statement is represented in Fig. 3.4.

Wikidata supports two special cases of statements for clarifying that the property has no value at all and the value of property is unknown. These two situations differ from the incompleteness of the statement. For example, one statement clarifies that Bruce Lee's date of birth is unknown, and another statement says that James Buchanan has no wife at all. In the first case, the statement indicates that the value of property is unknown, and in the second case, it clarifies that the property has no value at all. The statements are associated with a set of references which support their correctness. A reference, depending on its type, can be represented in different ways. Wikidata utilizes different external identifiers like the International Standard Name Identifier (ISNI) and authority files. Other key features of Wikidata are as follows:

- Wikidata covers entity labels, aliases, and descriptions approximately in all languages (about 350 languages),
- It can be queried with some un-official existing SPARQL endpoints,
- It is accessible online via HTTP lookup and supporting data format of RDF, JSON, XML, and SQL,
- It links to Freebase,
- It included textual auxiliary information of relations,
- It has a fixed schema which can be extended.

3.2.1.3 DBpedia

DBpedia (Lehmann et al., 2015) is a multilingual knowledge graph gathered from Wikipedia. DBpedia supports 111 languages and is collected from structured data of Wikipedia including infoboxes, lists, and tables. The user community of DBpedia

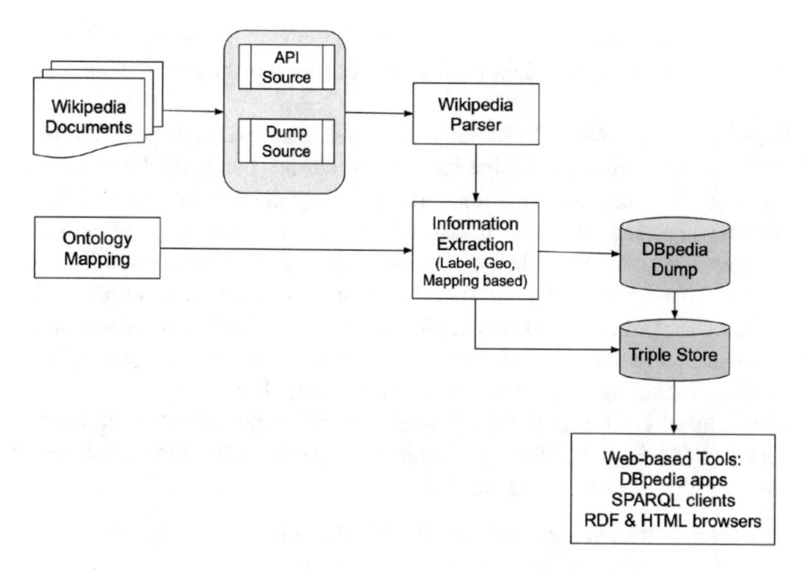

Fig. 3.5 The architecture of the extraction framework used by DBpedia

determines how to map the content of Wikipedia pages to the ontology of DBpedia. Wikipedia pages include various types of data beside the text data including infobox, categorized information, images, links to other pages, and geographical coordinates. The knowledge extraction framework of DBpedia extracts this information and stores it in a knowledge graph. The architecture of the extraction framework of DBpedia is shown in Fig. 3.5. The extraction framework receives the Wikipedia pages from the MediaWiki API or reads it from the dump of Wikipedia. In the parsing component, the input Wikipedia page is converted to an abstract syntax tree (AST). The AST is passed to several extractors for retrieving various information like labels and abstracts. Given an AST, each extractor generates a set of RDFs. Finally, outputs of extractors are stored in a sink.

Twenty-four different extractors are utilized for converting each part of Wikipedia pages into RDF statements. The available extractors are classified into four major categories including (1) Mapping-based Infobox Extraction, (2) Raw Infobox Extraction, (3) Feature Extraction, and (4) Statistical Extraction. The information is stored in the Wikipedia infoboxes in the format of attribute-value pairs. Various templates are used by users for filling the infoboxes in Wikipedia pages which results in inconsistency in describing the same objects or types. For example, the place of birth is represented with *"birthplace"* and *"placeofbirth"* attributes, and there is no unique name for describing this attribute. In the Raw Infobox Extraction method, the infobox content is directly mapped to RDF, and due to the mentioned problem, it has lower quality. In the Mapping-based Infobox Extraction method, a set of mappings are created by community to map the objects with the same concept and different names to a unique name in DBpedia ontology. The Feature Extraction method retrieve features from input article, and the Statistical Extraction method gathers statistics information about different pages and merges them. The statistical information can be word count or source of page links.

DBpedia covers a hierarchy of 320 classes that are explained by 1650 properties. DBpedia is collected from Wikipedia articles that are being modified or changed consistently. The frequent modifications of Wikipedia articles make data stored in DBpedia outdated and invalid. To mitigate the mentioned problem and keep DBpedia updated, the DBpedia live system is developed which captures the updates and changes from Wikipedia and modifies DBpedia based on them. By following the Linked Data principles, DBpedia includes a set of outgoing links which refer to other datasets and a set of incoming links from other datasets mentioning DBpedia. Count of the incoming links to DBpedia is about 39,007,478 links.

Users can access the data of Wikipedia by executing SPARQL queries or by downloading the datasets where each dataset is gathered by one of the extractors. The key features of DBpedia are as follows:

- DBpedia is collected automatically from Wikipedia,
- It is Linked to other datasets including Freebase and OpenCyc,
- It stores facts in the form of triples without storing any additional metadata,
- It has two static and live versions where the static version is updated once a year,
- It provides online executing of SPARQL queries,
- Ii includes human-readable IDs and classes and describes the entities by properties.

3.2.1.4 YAGO

YAGO is a joint project of Max Planck Institute and Telecom ParisTech University since 2007 (Suchanek et al., 2008). YAGO is another semantic knowledge base collected from diverse sources including Wikipedia (categories, infoboxes, redirects), WordNet (hyponymy, synset), Wikidata, GeoNames, etc. YAGO has been published in several versions. The latest version of YAGO includes more than 120 million facts describing more than 10 million entities. YAGO enhances the facts by adding temporal and spatial features to them. Each fact is stored as SPOTL where S is subject, P is predicate, O is object, T is time, and L is location.

YAGO covers Wikipedia pages from ten different languages. Accuracy of YAGO, which is evaluated manually by sampling the facts, is about 95%. YAGO is accessible by SPARQL queries in online browsers, and it can be downloaded directly in both .tsv and .ttl (RDF) formats from its Web page.[1] The source code of YAGO can be found on GitHub. No additional description of entities is stored in YAGO, while a description of relations is provided. YAGO has human-readable IDs and contains simple data types. YAGO has a link to DBpedia.

[1] Download YAGO from https://www.mpi-inf.mpg.de/departments/databases-and-information-systems/research/yago-naga/yago/downloads.

3.2.2 Datasets

3.2.2.1 WebQuestions

For gathering the WebQuestions dataset (Berant et al., 2013), the questions including only one entity and beginning with wh-question words are collected by using Google Suggest API. The breadth-first search algorithm is performed over questions beginning with question "Where was Barack Obama born?". The question is queried by eliminating the entity and the phrase describing the entity before or after the entity. Five candidate questions are produced for each query and appended to the queue. The procedure is continued until generating 1M questions and 100K questions are chosen randomly and submitted to Amazon Mechanical Turk (MTurk).

The generated questions are answered by workers of MTurk using the Freebase knowledge graph. The workers are asked to answer each question by the Freebase page of its entity. The answers must be selected from entities, values, or a list of entities in the page. The questions with no answer in Freebase are marked as unanswerable. At least 2 workers agreed on the answer of 6642 questions. Entities are mapped according to a Lucene index which is built over the Freebase entities. The dataset is divided into train and validation sets with 80% and 20% size of the collected data, respectively.

Sometimes the answers provided by workers are not correct due to the incompleteness of Freebase. In another case, the incompleteness of Freebase results in incomplete answers. For example, consider a question asking about a list of movies, which a specific actor played in them and Freebase includes half of the movies. In this case, the answer returned by workers is partial.

3.2.2.2 SimpleQuestions

The SimpleQuestions dataset (Bordes et al., 2015) is collected according to the facts of Freebase. SimpleQuestions includes 108,442 questions asked in natural language by English speaker annotators. The collected questions are divided into training, validation, and test sets with 75,910 (70%), 10,845 (10%), and 21,687 (20%) questions. Each fact is represented with a triple (subject, relation, object), and the goal of KBQA is defined as finding and comprehending the subject and relation from question sentences and retrieving the object from the knowledge graph. The SimpleQuestions dataset is collected by retrieving the facts and forming the question given the subject and relation.

SimpleQuestions is collected in two steps. In the first step, a subset of facts are selected from Freebase, and facts with undefined relation are eliminated. The facts with more than a specific number of objects for the given subject and relation are removed to avoid proposing questions like " Name a person who is an actor?". Examples of questions with corresponding facts are shown in Table 3.7.

Table 3.7 Example of questions from SimpleQuestions dataset and related facts from Freebase knowledge graph

Question	Related facts
What American cartoonist is the creator of Andy Lippincott?	(andy_lippincott, character_created_by, garry_trudeau)
Which forest is Fires Creek in?	(fires_creek, containedby, nantahala_national_forest)
What does Jimmy Neutron do?	(jimmy_neutron, fictional_character_occupation, inventor)

Table 3.8 Statistics of two subsets of SimpleQuestions and WebQuestions datasets (Bordes et al., 2015)

	WebQuestions	SimpleQuestions
Train	3000	75,910
Validation	778	10,845
Test	2032	21,687

In the second step, the facts retrieved in the previous step are sampled and given to workers for generating questions. To guarantee variability of the questions, the frequency of relationships is calculated, and the facts including a relationship with lower probability are sampled. Then, the sampled facts with their link to Freebase are given to human annotators for proposing the questions. The proposed questions must contain the subject entity and the relationship with the object as answer. Statistics of SimpleQuestions and WebQuestions datasets are provided in Table 3.8.

3.2.2.3 LC-QuAD 2.0

The LC-QuAD dataset (Dubey et al., 2019) is collected automatically. The process of generating the questions is represented in Fig. 3.6. The process starts with extracting a set of appropriate entities for generating questions. The entities are selected according to the Wikipedia Vital article.[2] The related same-as-links mentioning the Wikipedia IDs are discovered. For instance, the entity "Barack Obama" is an entity from the list of entities.

In the second step, the templates for SPARQL query are generated in a way that they cover different types of questions. The knowledge base architectures and the QA datasets are used for creating the SPARQL templates. The generated questions can have ten different types, including single-fact, multi-fact, count, Boolean, ranking, and temporal aspect.

In the third step, a list of suitable predicates for each question type is extracted because each question type supports a specific set of predicates for question generation. For example, a question with the "Count" type cannot support a

[2] https://en.wikipedia.org/wiki/Wikipedia:Vital_articles/Level/5.

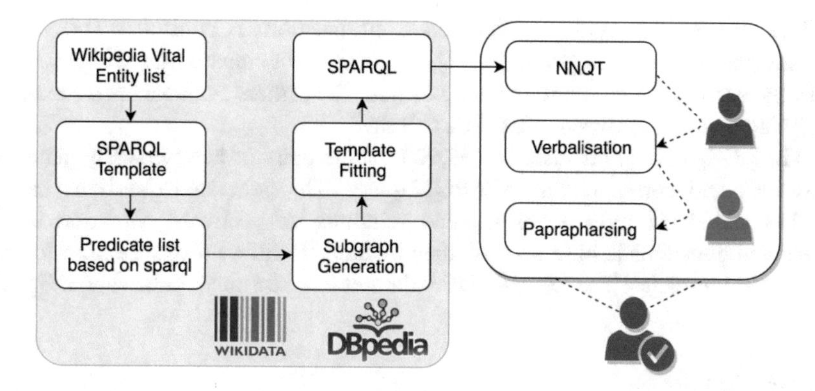

Fig. 3.6 The process of generating the LC-QuAD 2.0 dataset (Dubey et al., 2019) (" ©Springer Nature Switzerland AG 2019, reprinted with permission")

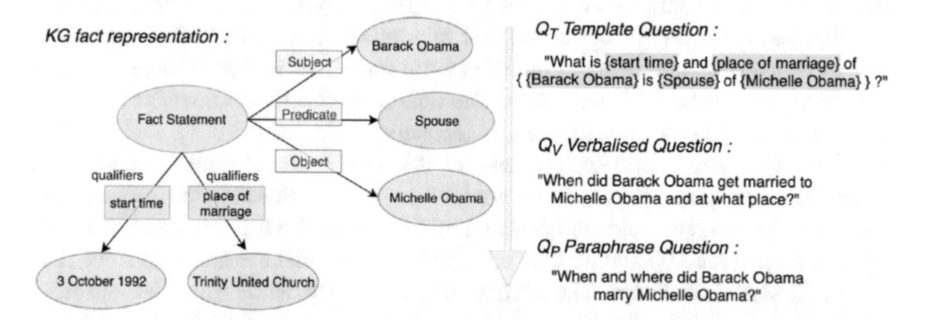

Fig. 3.7 An example of a subgraph generated for question generation with (right) the corresponding question template, verbalized question, and paraphrase question (Dubey et al., 2019) (" ©Springer Nature Switzerland AG 2019, reprinted with permission")

predicate like "birthPlace." After the third step, a subgraph is generated from the knowledge graph based on the entity, SPARQL template, and predicate list. An example of a subgraph is shown in Fig. 3.7. Then, the SPARQL query is generated and converted to a natural language template named question type (Q_T).

In the next steps, MTurk is utilized for creating natural language questions from the templates generated by the system. MTurk is perfumed in three steps. In the first step of the MTurk process, the task of verbalization is performed to convert the question template to its verbalized form $Q_T \rightarrow Q_V$. This step is required because the question template Q_T usually has grammatical errors and is semantically inconsistent. A set of instructions are defined for each question type and are given to the turkers to perform the first MTurk step. In the second step of the MTurk process , the turkers are responsible for paraphrasing the generated question $Q_V \rightarrow Q_P$ by altering the structure or syntax of the Q_V, while the semantic meaning of the Q_V is preserved. Turkers can perform the paraphrasing by using synonyms and aliases and altering the syntactic structure. In the third step of the

MTurk process, the quality of the generated questions is controlled and verified by humans. Turkers are given Q_T, Q_P, and Q_V to compare and decide whether they have the same semantic meaning or not. The turkers evaluate each sample by selecting a choice between "Yes/No/Can't say."

The LC-QuAD 2.0 dataset has 30,000 unique pairs of automatically generated questions and corresponding SPARQL query. The generated questions include 21,258 and 1310 unique entities and relations, respectively. Variation of the generated questions is high because they include 10 different types extracted from 22 unique templates. The LC-QuAD 2.0 dataset is based on Wikidata and DBpedia.

3.2.2.4 QALD

The goal of the QA over Linked Data (QALD) challenge (Usbeck et al., 2017) is to provide a benchmark for evaluating four different tasks. The main task of QALD is defined as returning the correct answer of a question or a SPARQL query which can extract the answer of input question from one or several RDF dataset(s) and supplementary knowledge resources. The data is stored in the format of QALD-JSON and can be downloaded from its repository.[3] One of the tasks supported by QALD is QA over Wikidata. The goal of this task is to measure the adaptation ability of the proposed models to unseen data sources. To this aim, the questions are generated for DBpedia and are answered by Wikidata. The training and test sets of the dataset include 100 and 50 questions, respectively, that are gathered from Task 1 of the QALD-6 challenge. The queries are generated to answer the questions from Wikidata.

3.2.3 Metrics

Some of the questions in KBQA are answered by a single fact, while some of them are list questions, and their answer is a list of entities retrieved from the knowledge source(s). The accuracy metric is used for evaluating the single-fact questions, and other metrics including the precision, recall, and F1-score metrics are mostly used for the evaluation of list questions. As the accuracy metric has been explained in Sect. 3.1.2, we won't explain it again. The other metrics are explained below.

- **F1-score:** Performance of the QA datasets which include the question with a set of correct answers is measured using the F1-score. F1-score is calculated by incorporating the precision and recall metrics.

  Precision measures the ability of the system in answering the questions correctly, while recall measures the ability of the system in retrieving all of the

[3] https://github.com/ag-sc/QALD/tree/master/7/data.

correct answers. Precision is defined as the portion of correct retrieved answers to all of the retrieved answers. Recall is calculated by dividing the number of correct answers to all of the existing correct answers. The two metrics are calculated as follows:

$$\text{Precision} = \frac{\text{number of correct retrieved answers}}{\text{number of retrieved answers}} \tag{3.9}$$

$$\text{Recall} = \frac{\text{number of correct retrieved answers}}{\text{number of existing correct answers}} \tag{3.10}$$

The F1-score is the harmonic mean of precision and recall metrics which pays attention to both objectives: (1) retrieving all of the correct answers and (2) correctness of the retrieved answers. Considering uniform weight for precision and recall, F1-score is calculated as follows:

$$\text{F1-score} = 2 * \frac{(\text{Precision} * \text{recall})}{\text{Precision} + \text{recall})} \tag{3.11}$$

Consider the following question from WebQuestions dataset with an array of nine correct answers in the gold dataset which are extracted from Freebase:
Question: *"what movies has Carmen Electra been in?"*
Answer: ["The Mating Habits of the Earthbound Human," "Scary Movie," "Getting Played," "Cheaper by the Dozen 2," "Meet the Spartans," "I Want Candy," "Full of It," "The Chosen One: Legend of the Raven," "Scary Movie 4," "Dirty Love"]

A model retrieves a set of the 12 answers for the given question as follows:

["How the Grinch Stole Christmas!", "All About Us," **"Scary Movie,"** "10 Items or Less," **"Getting Played," "Meet the Spartans," "I Want Candy," "Full of It,"** "Bruce Almighty," "Batman Begins," **"Dirty Love,"** "A Raisin in the Sun"]

The retrieved answer contains six correct answers and six incorrect answers. In this example, precision is $\frac{6}{12} = 0.5$, since 6 out of 12 retrieved answers are correct; and recall is $\frac{6}{9} = 0.66$, since 6 out of 9 correct answers are retrieved. Having the precision and recall values, F1-score is calculated as follows:

$$\text{F1-score} = 2 * \frac{(0.5 * 0.66)}{(0.5 + 0.66)} = 0.57 \tag{3.12}$$

3.3 Summary

We studied the evaluation of TextQA and KBQA in Sects. 3.1 and 3.2, respectively. In Sect. 3.1, we introduced the well-known datasets collected for the evaluation of TextQA and explained the metrics used for evaluation. We devoted Sect. 3.2 to introduce the knowledge bases, datasets, and metrics used for the evaluation of KBQA.

References

Abbasiantaeb, Z., & Momtazi, S. (2021). Text-based question answering from information retrieval and deep neural network perspectives: A survey. *Wiley Interdisciplinary Reviews: Data Mining and Knowledge Discovery*, e1412.

Berant, J., Chou, A., Frostig, R., & Liang, P. (2013). Semantic parsing on Freebase from question-answer pairs. In *Proceedings of the 2013 Conference on Empirical Methods in Natural Language Processing*, Seattle, Washington, USA (pp. 1533–1544). Association for Computational Linguistics. https://www.aclweb.org/anthology/D13-1160.

Berners-Lee, T., Hendler, J., & Lassila, O. (2001). The semantic web. *Scientific American, 284*(5), 34–43.

Bordes, A., Usunier, N., Chopra, S., & Weston, J. (2015). Large-scale simple question answering with memory networks. Preprint. arXiv:1506.02075.

Diefenbach, D., Lopez, V., Singh, K., & Maret, P. (2018). Core techniques of question answering systems over knowledge bases: a survey. *Knowledge and Information Systems, 55*(3), 529–569. ISSN:0219-3116.

Dubey, M., Banerjee, D., Abdelkawi, A., & Lehmann, J. (2019). Lc-quad 2.0: A large dataset for complex question answering over wikidata and dbpedia. In *International Semantic Web Conference* (pp. 69–78). Springer.

Feng, M., Xiang, B., Glass, M. R., Wang, L., & Zhou, B. (2015). Applying deep learning to answer selection: A study and an open task. *2015 IEEE Workshop on Automatic Speech Recognition and Understanding (ASRU)* (pp. 813–820).

Freedman, G., & Reynolds, E. G. (1980). Enriching basal reader lessons with semantic webbing. *The Reading Teacher, 33*(6), 677–684.

Guns, R. (2013). Tracing the origins of the semantic web. *Journal of the American Society for Information Science and Technology, 64*(10), 2173–2181.

Lehmann, J., Isele, R., Jakob, M., Jentzsch, A., Kontokostas, D., Mendes, P. N., Hellmann, S., Morsey, M., Van Kleef, P., Auer, S., et al. (2015). Dbpedia–a large-scale, multilingual knowledge base extracted from wikipedia. *Semantic Web, 6*(2), 167–195.

Suchanek, F. M., Kasneci, G., & Weikum, G. (2008). Yago: A large ontology from wikipedia and wordnet. *Journal of Web Semantics, 6*(3), 203–217.

Tapaswi, M., Zhu, Y., Stiefelhagen, R., Torralba, A., Urtasun, R., & Fidler, S. (2016). Movieqa: Understanding stories in movies through question-answering. In *2016 IEEE Conference on Computer Vision and Pattern Recognition (CVPR)* (pp. 4631–4640).

Usbeck, R., Ngomo, A.-C. N., Haarmann, B., Krithara, A., Röder, M., & Napolitano, G. (2017). 7th open challenge on question answering over linked data (qald-7). In *Semantic Web Evaluation Challenge* (pp. 59–69). Springer.

Vrandečić, D., & Krötzsch, M. (2014). Wikidata: a free collaborative knowledgebase. *Communications of the ACM, 57*(10), 78–85.

Wan, S., Lan, Y., Guo, J., Xu, J., Pang, L., & Cheng, X. (2016a). A deep architecture for semantic matching with multiple positional sentence representations. In *Proceedings of the Thirtieth AAAI Conference on Artificial Intelligence, AAAI'16* (pp. 2835–2841). AAAI Press.

Wan, S., Lan, Y., Xu, J., Guo, J., Pang, L., & Cheng, X. (2016b). Match-srnn: Modeling the recursive matching structure with spatial rnn. In *IJCAI*.

Yang, Y., Yih, W.-t., & Meek, C. (2015). Wikiqa: A challenge dataset for open-domain question answering. In *Proceedings of the 2015 Conference on Empirical Methods in Natural Language Processing*, Lisbon, Portugal (pp. 2013–2018). Association for Computational Linguistics. https://doi.org/10.18653/v1/D15-1237.

Chapter 4
Introduction to Neural Networks

Abstract Since deep learning models include a wide range of the recent models in both TextQA and KBQA, we introduce the main neural models used in this field in this chapter. We study the neural architectures widely used in Sect. 2.3.2 and avoid repeating some of the details described in this section in the following chapters. We also describe the available word representation model from traditional and state-of-the-art models that are utilized in QA systems.

4.1 Neural Architectures

Neural networks are developed for solving classification and regression problems. The architecture of neural networks is inspired by the human brain. Our brain is capable of performing many perceptual acts that demand computation. The human brain contains more than 10 billion connected neurons that are information processing units of our brain. Each neuron receives impulses from other neurons and carries the information to other cells.

Like our brain, the building blocks of artificial neural networks are nodes (neurons), which are basic units of computations. The architecture of a neural network is demonstrated in Fig. 4.1. Neural networks include input, output, and hidden layers, activation function, and learning rules. Each layer is a group of nodes and layers are connected by connection weights. The input nodes are used to carry the information to the next layer, and no computation is performed in them. Computation is performed in the hidden layers by connections. The connections connect the output of a layer to input of another layer, and they are assigned weights. In the output nodes, an activation function is applied to map the result of computations to the required format for output. The activation functions bring

S. Momtazi, Z. Abbasiantaeb, *Question Answering over Text and Knowledge Base*,
https://doi.org/10.1007/978-3-031-16552-8_4

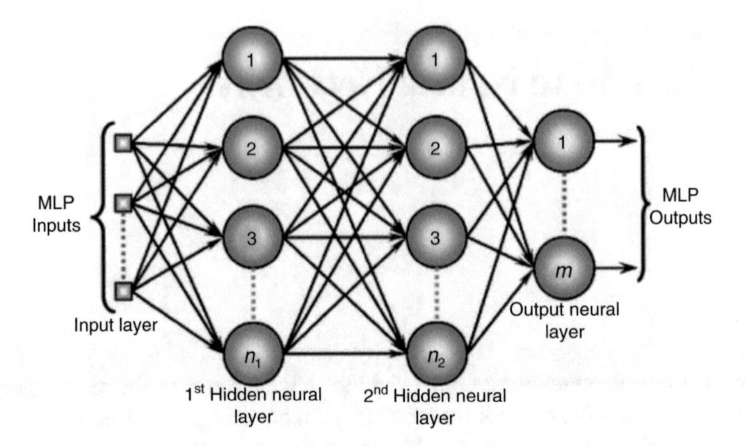

Fig. 4.1 The architecture of a neural network (da Silva et al., 2017) ("©2017 Springer International Publishing Switzerland, reprinted with permission")

non-linearity to neural computations. Some of the well-known activation functions are as follows:

- **Sigmoid:**

$$f(x) = \frac{1}{1 + e^{-x}} \tag{4.1}$$

- **Tanh:**

$$f(x) = \tanh(x) = \frac{2}{1 + e^{-2x}} - 1 \tag{4.2}$$

- **ReLU:**

$$f(x) = \begin{cases} 0 & \text{for } x < 0 \\ x & \text{for } x \geq 0 \end{cases} \tag{4.3}$$

- **SoftMax**

$$\sigma\left(x_j\right) = \frac{e^{x_j}}{\sum_i e^{x_i}} \tag{4.4}$$

The parameters of a neural model are learned through a learning process according to the learning rule. In other words, the learning rule modifies the connection weights or parameters to learn the function which maps the input to output. The learning process, which is performed by updating the connection weights, is called training.

4.1.1 Feed-Forward Neural Network

In feed-forward neural networks, information travels in the forward direction from input layer to output layer, and no cycle is created. Feed-forward neural networks can be single-layer or multi-layer. In the single-layer neural network, the input layer is connected to the output layer directly without any hidden units, whereas the multi-layer neural network contains one or several hidden layers. The architecture of a multi-layer neural network with two hidden layers is shown in Fig. 4.1. The learning process is conducted by computing an error function (or cost function) which calculates the difference between the observed value of output nodes and their expected value. Different optimization algorithms are available for optimizing the neural weights including gradient descent (GD), stochastic gradient descent (SGD), adaptive gradient (AdaGrad), RMSprop, and Adam. The weights and biases are initialized randomly, and the learning process updates them by minimizing the error value of nodes. By training the neural model, the connection weights are determined to solve the regression or classification problem. An optimization algorithm is used for minimizing the cost function over neural architecture.

In a fully connected network, each node in the hidden layers and output layer is connected to all of the nodes from the previous layer with a specific connection weight. The connections are in the forward direction which connect a start node to an end node from the next layer. The value of each node is calculated based on its input connections and the values of corresponding start nodes from the previous layer. Weighted summation of start nodes from the previous layer with weight of corresponding connections plus bias term determines the value of each node. An example of calculating the node values in neural networks is shown in Fig. 4.2. This neural network has two hidden layers with three and two nodes, respectively, and an activation function is applied to all of the nodes. According to this figure, the value

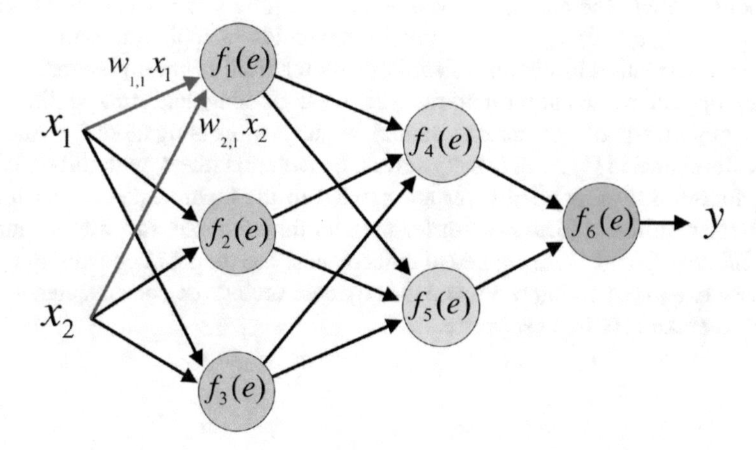

Fig. 4.2 An example of calculating the node values in a feed-forward neural network

of the first node in the first hidden layer y_1 is calculated as follows:

$$y_1 = f_1(w_{1,1}x_1 + w_{2,1}x_2) \tag{4.5}$$

where f_1 represents the activation function, $w_{1,1}$ is the weight of the connection connecting the first node of input layer x_1 to the first node of the second layer y_1, and $w_{2,1}$ indicates the weight of the connection between the second node of the input layer x_2 and the first node of the second layer y_1. In the case of having bias term b, the above formulation is changed to the following equation:

$$y_1 = f_1(w_{1,1}x_1 + w_{2,1}x_2 + b) \tag{4.6}$$

4.1.2 Convolutional Neural Network (CNN)

CNNs are used for sentence modeling by extracting n-gram features of the input sentence (Young et al., 2018). CNNs include several layers of convolution followed by a pooling layer. The input sentence is passed to the convolution layer by an embedding matrix $W \in \mathcal{R}^{n \times d}$. Row i of matrix W is embedding of the i-th word within the input sentence. Each word is represented by a d-dimensional vector embedding $w_i \in \mathcal{R}^d$. The same convolution function (filter) is applied to each window of h consecutive words $w_{i:i+h-1}$ to produce feature map c for each n-gram as follows:

$$c_i = f\left(w_{i:i+h-1} \cdot k^T + b\right) \tag{4.7}$$

where b is the bias, f represents a non-linear activation function, and $k \in \mathcal{R}^{hd}$ is the convolution filter. The output of convolution with filter k is a sequence of feature maps $c = [c_1, c_2, \ldots, c_{n-h+1}]$. A pooling operation is applied to output feature maps of convolution to obtain a fixed-size vector from n-gram patterns. Various pooling operations including max-pooling, mean-pooling, and min-pooling exist, and usually max-pooling is used in natural language processing tasks $\hat{c} = \max\{c\}$.

As it is shown in Fig. 4.3, usually several convolution filters, with different filter sizes, followed by a pooling layer are applied to the input sequence. Each filter size extracts different local n-gram features. In this example, two filters with two different sizes (3 and 4) are applied to the sentence. The output of convolution filters is passed to a max-pooling layer, and the obtained vectors are concatenated to form the representation of input sentence.

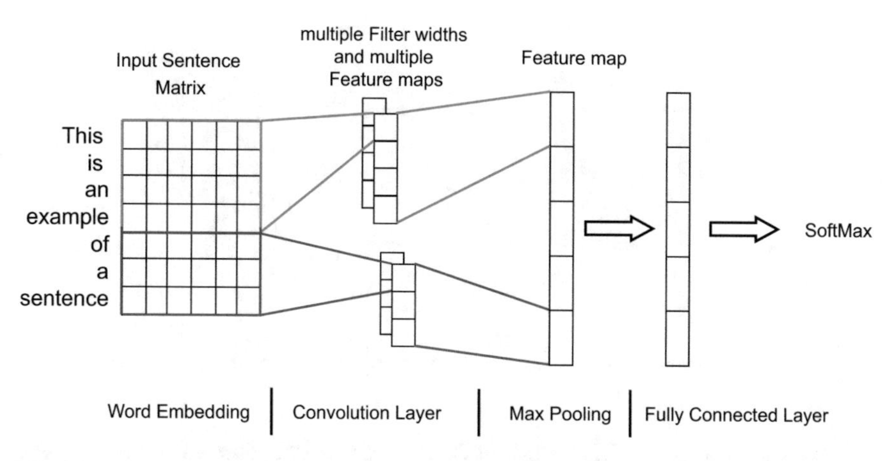

Fig. 4.3 An example of CNN for sentence modeling (Ferjani et al., 2019) ("©2019 Springer Nature Switzerland AG, reprinted with permission")

4.1.3 Recurrent Neural Network (RNN)

Recurrent neural networks (RNNs) are capable of remembering and utilizing information of the past inputs via feedback connection, and this feature makes them a proper choice for processing the sequential and time-series data (Zhang et al., 2018). The textual data is sequential because the sentences generated in languages are made up of a sequence of words that appear in a specific order such that their spatial dependency builds up the meaning of the sentence. As a result, RNNs are widely applied in various natural language processing tasks and have shown great performance. RNNs take the input of the current time step (x_t) and the hidden state of the previous time step (h_{t-1}) to generate the hidden state of the current time step (h_t) in their output. The output of RNN is a sequence of hidden states which are considered as contextual representation of the input sequence up to that token. In other words, the hidden states are memory cells which carry the information of sequence up to each time step. A standard RNN computes the hidden state of time step t as follows:

$$h_t = \tanh\left(W_{xh}x_t + W_{hh}h_{t-1} + b_h\right) \tag{4.8}$$

where W_{ab} is the weight matrix and b_h is the bias. The tanh activation function can be replaced with other activation functions like *ReLU*. Standard RNN has a vanishing gradient problem which is solved in other variants of RNN by using gate mechanism.

In the following, we will introduce two variations of RNNs, namely, long short-term memory (LSTM) and gated recurrent units (GRU), which are mostly utilized in natural language processing tasks including QA.

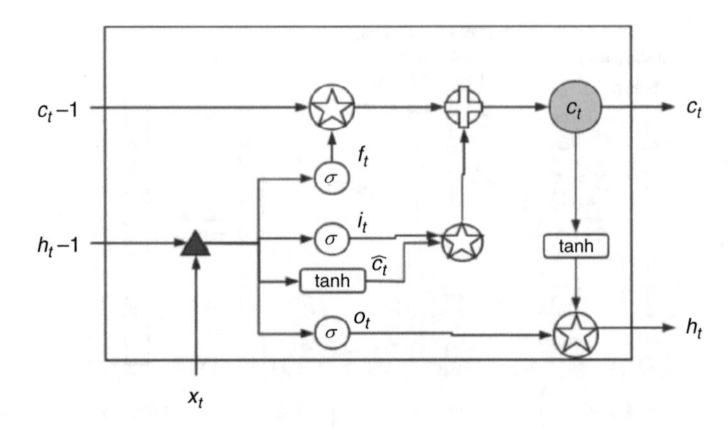

Fig. 4.4 The architecture of LSTM (Xie et al., 2022) ("©Springer Nature B.V, reprinted with permission")

- **Long Short-Term Memory (LSTM):** LSTM overcomes the vanishing gradient problem by using a gating mechanism. The gating mechanism controls the backpropagation of error to previous time steps. The architecture of LSTM is shown in Fig. 4.4. LSTM has three input, output, and forget gates and calculates the hidden state at each time step t as follows:

$$\begin{aligned} i_t &= \sigma\left(W_{xi}x_t + W_{hi}h_{t-1} + b_i\right) \\ f_t &= \sigma\left(W_{xf}x_t + W_{hf}h_{t-1} + b_f\right) \\ c_t &= f_t \odot c_{t-1} + i_t \odot \tanh\left(W_{xc}x_t + W_{hc}h_{t-1} + b_c\right) \\ o_t &= \sigma\left(W_{xo}x_t + W_{ho}h_{t-1} + b_o\right) \\ h_t &= o_t \odot \tanh\left(c_t\right) \end{aligned} \tag{4.9}$$

  where i, f, and o represent input, forget, and output gates, respectively, c is the cell unit, W_{ab} is the weight matrix to be learned, and $\odot$ is the element-wise multiplication.
- **Gated Recurrent Unit (GRU):** GRU controls the flow of information from previous states by the update z and the reset r gates. The architecture of GRU is shown in Fig. 4.5. The reset gate controls the amount of information from the previous hidden state h_{t-1} in calculating h'_t, and the update gate determines the amount of information passed from the previous hidden state h_{t-1} in calculating

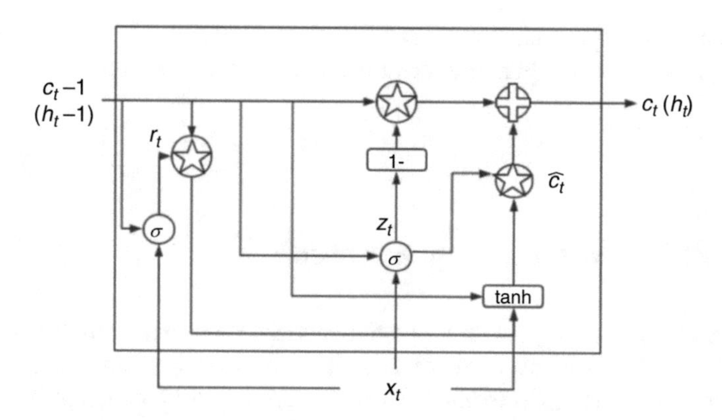

Fig. 4.5 The architecture of GRU (Xie et al., 2022) ("©Springer Nature B.V, reprinted with permission")

the current hidden state h_t. GRU has a less complicated architecture than LSTM. GRU calculates the hidden state h_t as follows:

$$
\begin{aligned}
r_t &= \sigma\left(W_{xr}x_t + W_{hr}h_{t-1} + b_r\right) \\
z_t &= \sigma\left(W_{xz}x_t + W_{hz}h_{t-1} + b_z\right) \\
h_t' &= \tanh\left(W_{xh}x_t + W_{hh}\left(r_t \odot h_{t-1}\right) + b_h\right) \\
h_t &= z_t \odot h_{t-1} + \left(1 - z_t\right) \odot h_t'
\end{aligned}
\tag{4.10}
$$

where W_{ab} is the model parameter to be learned and b_x is the bias.

Several layers of RNNs can be stacked by passing the output time steps of an RNN to another RNN. In this case, each layer extracts a different level of abstraction, and the output of the last RNN that is on top of the other RNNs is used.

As RNNs process the input sequence in one direction, each time step is only informed about the previous inputs and has no information about the future. The textual data mostly has bidirectional dependencies that means each word in one sentence has dependencies with not only the previous words but also the next words. Context of each word is defined as the window of w words before and after it. To benefit from both directions, BiRNNs are used in natural language processing tasks. In this case, two separate RNNs process the input sentence in the forward and the backward directions which results in producing two sequences of time steps as:

$$
\begin{aligned}
&[\overrightarrow{h_0}, \overrightarrow{h_1}, \cdots, \overrightarrow{h_n}] \\
&[\overleftarrow{h_0}, \overleftarrow{h_1}, \cdots, \overleftarrow{h_n}]
\end{aligned}
\tag{4.11}
$$

Then the final representation of each time step is generated by concatenating the hidden state of the corresponding forward and backward directions as follows:

$$h_t = [\overrightarrow{h_t}; \overleftarrow{h_t}] \tag{4.12}$$

4.2 Distributed Word Representation

By representing the words with a one-hot encoded representation, they are represented in a V-dimensional space which each dimension represents a specific word. In this space, no semantic relation between them could be captured because the words do not share any dimension with each other. Another problem of this model is the curse of dimensionality. To mitigate the mentioned problems, the idea of distributed word representation model by aid of neural networks is proposed.

Distributed word representation models create a representation for words based on their semantic and syntactic features and encode them in a low-dimensional space. The words near to each other in this space have similar semantic and syntactic features, and as a result, similarity or distance of words could be measured by using distance metrics. In the following, we will explain two distributed word representation models that are widely used in natural language processing tasks, specifically QA systems.

4.2.1 Word2Vec

Word2Vec (Mikolov et al., 2013) is a group of related neural-based models which are used for generating the representation of words. Word2Vec extracts features from text for representing the words. Each dimension of the extracted vector indicates a special feature, and each word has a unique vector representation. These features are extracted by a neural network, and the concept of each feature is not clear. Word2Vec models use a two-layer feed-forward neural network trained for preparing the context of words. For training the Word2Vec models, a large-size text corpus, which includes all the vocabulary words, is required. The input words are transformed into a new space within the neural architecture, and a new vector representation is generated for words. The words with similar context in the corpus are mapped to similar vector representations. In the following, we will introduce two Word2Vec models including Skip-gram and Continuous Bag of Words (CBOW).

- **CBOW:** CBOW is trained for predicting each word given its context. Context of each target word is defined as a window of w words before and after the target word. A raw corpus is required for training the CBOW model. The training samples are generated by moving along the text tokens and selecting each word as target and its surrounding words as context. As the architecture of the CBOW

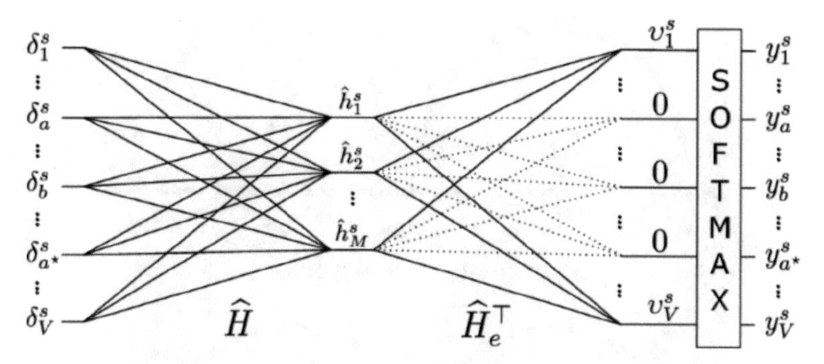

Fig. 4.6 The architecture of CBOW model (Di Gennaro et al., 2021) ("©Springer Science+Business Media, LLC, part of Springer Nature, reprinted with permission")

model is shown in Fig. 4.6, it consists of a simple fully connected network with one hidden layer. One-hot which encodes the representation of context words is passed to the input layer of CBOW, and the target word is predicted in the output layer. The input and output layers have V neurons where V is the vocabulary size and the hidden layer has M neurons. In the output layer, probability of being a target word is calculated for each word, and the word with the highest probability is selected as the target word. Input layer is connected to hidden layer with weight matrix $\hat{H} \in R^{V \times M}$, and the hidden layer is connected to the output layer by weight matrix $\hat{H}_e^T \in R^{M \times V}$. The two weight matrices are considered as word embeddings where the k-th row of the matrix $\hat{H}$ and k-th column of the matrix $\hat{H}_e^T$ indicate the representation of the k-th word of the vocabulary. Consequently, two vector embeddings v_c and v_w are generated for each word of vocabulary as follows:

$$v_c = \hat{H}_{(k,0)}, v_w = \hat{H}_e^T(0, k) \tag{4.13}$$

Input of the CBOW model is the summation of the one-hot encoding of all context words, and the hidden layer is generated as:

$$h = W^T (x_1 + x_2 + \ldots + x_c) \tag{4.14}$$

where c is the size of the context window.

- **Skip-gram:** The idea behind this model is that words happening in the same context can be replaced and they tend to be similar in semantics. The objective function is designed to map the words that appear in the similar context to near vector representations in the embedding space. The Skip-gram model has the same architecture as CBOW, but its training objective is different. Rather than predicting the target word, it predicts the context words given

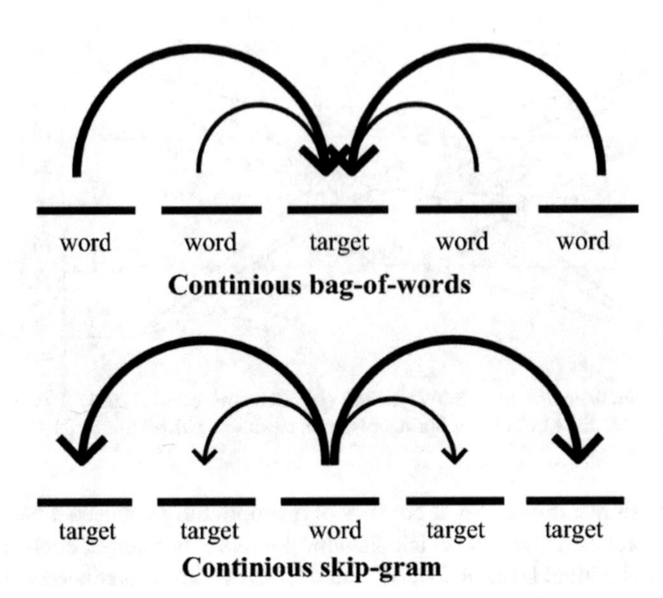

Fig. 4.7 The architecture of Skip-gram and CBOW model. The Skip-gram model predicts the context given the target word, while the CBOW model predicts the target word given the context (Hadifar & Momtazi, 2018) ("©Springer Nature B.V., reprinted with permission")

the target word. For example, when the size of context window is $C = 3$ given the word w_t, the model predicts other context words in the sequence $[w_{t-3}, w_{t-2}, w_{t-1}, w_t, w_{t+1}, w_{t+2}, w_{t+3}]$. The architecture of the Skip-gram model in comparison with the CBOW model is shown in Fig. 4.7. One-hot which encodes the representation of the target word is passed to the input layer, and in the output, the context words are predicted. Two input and output embedding matrices are generated in the hidden layer weights for representing the words. The model predicts $P(w_{t+j}|w_t)$ in the window of context words and is defined as below:

$$J(\theta) = \frac{1}{T} \sum_{t=1}^{T} \sum_{-C \leq j \leq C, j \neq 0} \log p\left(w_{t+j} \mid w_t\right) \tag{4.15}$$

where θ is all variables to be optimized.

4.2.2 *GloVe*

GloVe word embedding model (Pennington et al., 2014) creates the representation of words according to their co-occurrence in language. A word-word matrix X in which cell x_{ij} indicates the number of times word x_j appears in the context of word

x_i is used as the model input. The probability of observing word j in the context of word i is defined as:

$$P_{ij} = P(j \mid i) = \frac{X_{ij}}{X_i} \tag{4.16}$$

where $X_i = \sum_k X_{ik}$.

The above probability is used for modeling the semantic relation of words. Consider three words i, j, and k which word k is more related to word i than word j. In this case, the ratio of the two probabilities $\frac{P_{ik}}{P_{jk}}$ tends to be larger than one. On the contrary, when word k is more related to word j than word i, the ratio of the two probabilities $\frac{P_{ik}}{P_{jk}}$ tends to be smaller than one. In another case, when word k is not related to none of the i and j words, the ratio of the two probabilities $\frac{P_{ik}}{P_{jk}}$ will be near one. The main intuition behind the GloVe word embedding model is using the co-occurrence ratios for learning vector embeddings. A function F is defined as follows:

$$F\left(w_i, w_j, \tilde{w}_k\right) = \frac{P_{ik}}{P_{jk}} \tag{4.17}$$

where $w_i \in \mathbb{R}^d$ and $w_j \in \mathbb{R}^d$ represent the word vectors and $\tilde{w} \in \mathbb{R}^d$ indicates the context word vectors.

The function F should represent the properties of ratio in the embedding space. To this aim, the function F should work with difference of two input vectors as follows:

$$F\left(w_i - w_j, \tilde{w}_k\right) = \frac{P_{ik}}{P_{jk}} \tag{4.18}$$

Inputs of the function F are vectors, while the ratio of probabilities is a scalar value. To preserve the linear structure of function F, input is changed as:

$$F\left(\left(w_i - w_j\right)^T \tilde{w}_k\right) = \frac{P_{ik}}{P_{jk}} \tag{4.19}$$

The nature of the context and word in the word-word matrix is similar. According to this fact, the function F must be invariant to replacing the context with word $w \leftrightarrow \tilde{w}$ and $X \leftrightarrow X^T$. Consequently, the function F is defined as follows:

$$\begin{aligned} F\left(\left(w_i - w_j\right)^T \tilde{w}_k\right) &= \frac{F\left(w_i^T \tilde{w}_k\right)}{F\left(w_j^T \tilde{w}_k\right)} \\ F\left(w_i^T \tilde{w}_k\right) &= P_{ik} = \frac{X_{ik}}{X_i} \\ w_i^T \tilde{w}_k &= \log\left(P_{ik}\right) = \log\left(X_{ik}\right) - \log\left(X_i\right) \end{aligned} \tag{4.20}$$

The term $\log(X_i)$ is replaced with the bias term b_i as it does not depend on k and the bias term b_k is added to preserve the symmetry property as follows:

$$w_i^T \tilde{w}_k + b_i + \tilde{b}_k = \log(1 + X_{ik}) \tag{4.21}$$

A shift is made to the log function to prevent the infinite value of log near the zero. The above formula devotes equal weight to all co-occurrences which is not fair as some of the co-occurrences are rare and bring noise to the model. To overcome the mentioned issue, a weighted version of the cost function is defined as follows:

$$J = \sum_{i,j=1}^{V} f\left(X_{ij}\right)\left(w_i^T \tilde{w}_j + b_i + \tilde{b}_j - \log X_{ij}\right)^2 \tag{4.22}$$

where $f(X_{ij})$ indicates the weight of co-occurrence X_{ij}.

The function $f(X_{ij})$ should preserve some properties including (1) cost function must tend to zero when X_{ij} approaches zero, or $\lim_{x\to 0} f(x)\log^2 x$ must not be infinite, (2) weight function must give lower weight to rare co-occurrences, and (3) frequent co-occurrences must be assigned a small value to prevent overweighting. The following function F which maintains the mentioned conditions is proposed:

$$f(x) = \begin{cases} (x/x_{\max})^{\alpha} & \text{if } x < x_{\max} \\ 1 & \text{otherwise} \end{cases} \tag{4.23}$$

where $\alpha = 3/4$ and $x_{\max} = 100$.

The GloVe word embeddings are learned by optimizing Function 4.22.

4.3 Contextual Word Embedding

Traditional word representation models create a fixed encoding for each word regardless of the context in which they are appeared in. Some of the words have different meanings regarding their context and have polysemic behavior. For example, the word ‘bank’ can be a financial institute or the lang alone side the beach or mass when it is used in a sentence. Consequently, different meanings of word ’bank’ won’t be distinguished in its embedding as its context is not involved in the process of embedding generation and different meanings of word ‘bank’ share the same encoding.

The contextual word embedding models are proposed for mitigating the described problem. They create word embedding according to the context sentence. A word with multiple meanings has multiple representations that are produced according to the contextual meaning of the input word. We will explain three well-known contextual embedding models in the following sections.

4.3.1 ELMo

ELMo (Peters et al., 2018) is a task-specific contextual word embedding model which utilizes two distinct forward and backward LSTMs for calculating the probability of input sequence. The forward LSTM computes the probability of each token based on previous tokens, and the backward LSTM computes the probability of each token according to the following tokens based on the following equations.

$$
\begin{aligned}
p(t_1, t_2, \ldots, t_N) &= \prod_{k=1}^{N} p(t_k \mid t_1, t_2, \ldots, t_{k-1}) \\
p(t_1, t_2, \ldots, t_N) &= \prod_{k=1}^{N} p(t_k \mid t_{k+1}, t_{k+2}, \ldots, t_N)
\end{aligned}
\tag{4.24}
$$

The backward LSTM is similar to the forward LSTM, while the input sequence is given in reverse order to it. A multi-layer LSTM generates a hidden state representation for token t_k in each layer given the word representations x_k^{LM}. By using both the forward and backward LSTMs, two sets of forward $\overrightarrow{h}_{k,j}^{LM}$ and backward $\overleftarrow{h}_{k,j}^{LM}$ hidden states are produced.

The combination of a forward and backward LSTM is called bidirectional language model (BiLM). A BiLM, with L-layer forward and backward LSTMs, produces output set R with length $2L + 1$ for each input token t_k as follows:

$$
\begin{aligned}
R_k &= \left\{ x_k^{LM}, \overrightarrow{h}_{k,j}^{LM}, \overleftarrow{h}_{k,j}^{LM} \mid j = 1, \ldots, L \right\} \\
&= \left\{ h_{k,j}^{LM} \mid j = 0, \ldots, L \right\}
\end{aligned}
\tag{4.25}
$$

where $h_{k,0}^{LM}$ is the input token embedding x_k^{LM} and $h_{k,j}^{LM}$ is the concatenation of $\overrightarrow{h}_{k,j}^{LM}$ and $\overleftarrow{h}_{k,j}^{LM}$.

To use R_k in other models, ELMo converts it to a single vector $\mathrm{E\,L\,M\,o}_k = E\left(R_k; \Theta_e\right)$. The conversion can be performed by just selecting the hidden state of last layer L as $E\left(R_k\right) = h_{k,L}^{LM}$ or using the following weighting function:

$$
\mathrm{ELMo}_k^{\mathrm{task}} = E\left(R_k; \Theta^{\mathrm{task}}\right) = \gamma^{\mathrm{task}} \sum_{j=0}^{L} s_j^{\mathrm{task}} h_{k,j}^{LM}
\tag{4.26}
$$

where s_j^{task} is the softMax normalized weight of the j-th layer and γ^{task} is the scale of vectors generated by ELMo.

Pre-trained BiLM can be easily integrated with deep neural models in natural language processing. Usually, all of the deep models include a common word representation layer in the first layer which allows to add ELMo representations easily. In some neural models, the ELMo representation is obtained from a pre-trained BiLM with freeze weights and added to word representation $[x_k; \text{ELMo}_k^{task}]$ to enhance it. In some neural architectures with RNNs, ELMo representation is added to output hidden states of RNN for leveraging it.

4.3.2 BERT

Bidirectional Encoder Representations from Transformers (BERT) (Devlin et al., 2019) is another contextualized language model representation model which considers both left-to-right and right-to-left directions as context. The existing pre-trained language models are divided into two main categories, namely, feature-based and fine-tuning. The feature-based models use the pre-trained representation as an additional feature in their own architecture, while the fine-tuning approaches use the same architecture of the pre-trained language model in downstream tasks and fine-tune its parameters. For example, ELMo is a feature-based pre-trained language model as it uses the pre-trained embeddings as extra features.

BERT is a fine-tuning language model and is pre-trained on large-size unlabeled corpora with two Masked Language Model (MLM) and Next Sentence Prediction (NSP) tasks. The BookCorpus (800M words) (Zhu et al., 2015) and English Wikipedia (2,500M words) are used for pre-training BERT. In MLM, a specific percentage of input sentence tokens are masked randomly, and they are predicted in the output layer. After selecting the tokens for masking, the selected token will be replaced by $[MASK]$ token with probability of 80%, with another random token with probability of 10%, and won't be changed with probability of 10%. The NSP task is used for capturing the relation of two input sentences. In NSP, two input sentences are jointly fed to the BERT model, and the contextual representation of each input token is created in the output layer.

The pre-trained BERT model can be used in other downstream natural language processing tasks by adding an additional task-specific layer to its output and fine-tuning it. Application of BERT in various natural language processing downstream tasks has demonstrated great improvement. BERT includes several layers of bidirectional transformer encoder. Multi-head self-attention mechanism is used in the transformer encoder that helps the model to attend the text from both left-to-right and right-to-left directions.

The architecture of BERT in both pre-training and fine-tuning steps is shown in Fig. 4.8. After pre-training the BERT with unlabeled text, the same architecture is initialized with pre-trained weights and fine-tuned for a downstream task. All of the parameters are fine-tuned in the fine-tuning step. A special symbol $[CLS]$ is inserted to the beginning of each input, and inputs are separated by $[SEP]$ symbol.

Input of BERT is created according to Fig. 4.9. Token embeddings are summed with segment and position embeddings. The position embedding incorporates the sequential nature of the input sentence, and the segment embedding distinguishes the two input texts. For example, in the case of using BERT for QA, two input texts

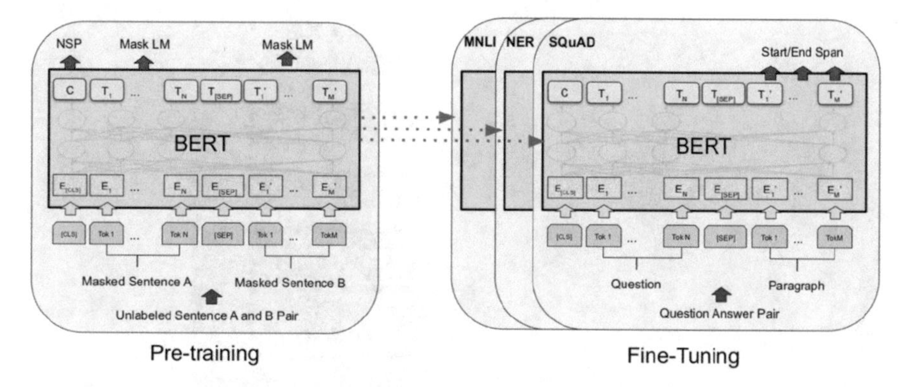

Fig. 4.8 Pre-training and fine-tuning steps in BERT (Devlin et al., 2019) ("©1963-2022 ACL, reprinted with permission")

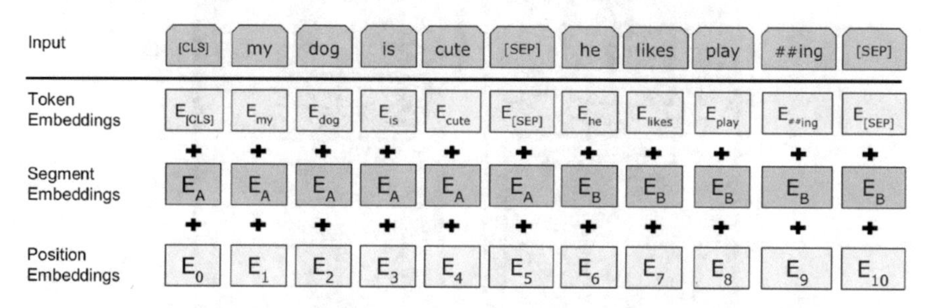

Fig. 4.9 Building input representation of BERT (Devlin et al., 2019) ("©1963-2022 ACL, reprinted with permission")

are a question and the candidate answer sentence, and they are separated by symbol $[SEP]$.

The transformer (Vaswani et al., 2017) has encoder-decoder architecture as is shown in Fig. 4.10, and the encoder of transformer is the main component of BERT. The transformer only relies on an attention mechanism for capturing the dependencies in sequence modeling tasks and has a stacked multi-layer architecture. Both the encoder and decoder parts shown in Fig. 4.10 are replicated for $N = 6$ times.

Input of transformer is a sequence of token representations $x = [x_1, \cdots, x_n]$, which the encoder encodes them to sequence of encoded representations $z = [z_1, \cdots, z_n]$. The encoded sequence is given to the decoder model which produces the sequence $y = [y_1, \cdots, y_m]$. The transformer's encoder and decoder consist of multi-head self-attention and position-wise feed-forward networks. According to Fig. 4.10, each encoder block includes one multi-head self-attention and one position-wise fully connected network which both are completed with residual connections. To be more specific, output of them is added to their input and then passed to a normalization layer: $\text{LayerNorm}(x + \text{Sublayer}(x))$. The decoder has an

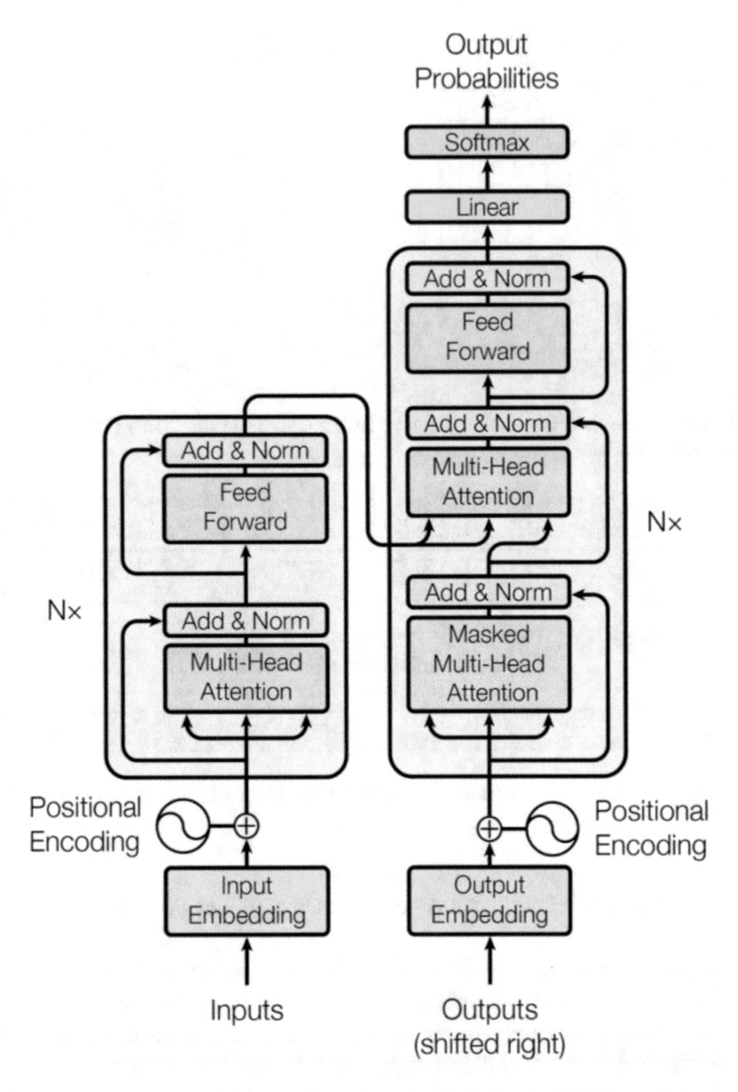

Fig. 4.10 Encoder-decoder architecture of transformer. The left part of the above figure represents the encoder, and the right part represents the decoder (Montenegro & da Costa, 2022) ("©Springer Nature Switzerland AG, reprinted with permission")

extra multi-head self-attention layer to incorporate the output of the encoder in the decoder.

In the self-attention mechanism, a set of queries (Q), keys (K), and values (V) are used as input to produce the attention output as follows:

$$\text{Attention}(Q, K, V) = \text{softMax}\left(\frac{QK^T}{\sqrt{d_k}}\right)V \tag{4.27}$$

where Q, K, and V are matrices of d_q, d_k, and d_v dimensional vectors, respectively. In the multi-head attention rather than applying just one self-attention layer, h self-attention layers are employed. The input queries, keys, and values are projected to h different matrices, and the projected matrices are passed to the self-attention layer in parallel. Finally, output of the self-attention layers is concatenated to form the output of multi-head attention as follows:

$$\begin{aligned}
&\text{MultiHead}(Q, K, V) = \text{Concat}\ (\text{head}_1, \ldots, \text{head}_\text{h})\, W^O \\
&\text{head}_\text{i} = \text{Attention}\left(QW_i^Q, KW_i^K, VW_i^V\right)
\end{aligned} \tag{4.28}$$

where head_i is the output of i-th self-attention and $W_i^Q \in \mathbb{R}^{d_{\text{model}} \times d_k}$, $W_i^K \in \mathbb{R}^{d_{\text{model}} \times d_k}$, $W_i^V \in \mathbb{R}^{d_{\text{model}} \times d_v}$, and $W^O \in \mathbb{R}^h d_v \times d_{\text{model}}$ are projection matrices.

A fully connected feed-forward layer is applied to each position in both the encoder and decoder models as follows:

$$\text{FFN}(x) = \max\,(0, xW_1 + b_1)\, W_2 + b_2 \tag{4.29}$$

This network includes two distinct linear transformations which are the same for various positions. Each layer has its own set of parameters.

4.3.3 RoBERTa

The architecture of Robustly Optimized BERT (RoBERTa) is the same as the BERT model. Similar to the architecture of BERT, RoBERTa consists of transformer encoder layers, but some modifications are applied in pre-training of RoBERTa. These modifications include (1) removing the NSP task in pre-training, (2) training with larger batch sizes, (3) using larger dataset for training, (4) dynamic masking, (5) removing NSP task, and (6) using longer sequences. RoBERTa has shown significant improvement in comparison to BERT in many downstream natural language processing tasks which indicates the effectiveness of modifications and capacity of BERT to improve.

4.4 Summary

We provided a brief explanation of neural networks in Sect. 4.1 to help the audience of the book in comprehending the neural models explained in the following sections. Due to the broad application of word representation models in QA models, we provided an explanation of them in Sect. 4.2. The contextualized word embeddings are widely used in various natural language processing tasks, and as they are used in QA models, we explained them in Sect. 4.3.

References

da Silva, I. N., Hernane Spatti, D., Andrade Flauzino, R., Liboni, L. H. B., & dos Reis Alves, S. F. (2017). *Multilayer perceptron networks* (pp. 55–115). Cham: Springer International Publishing. ISBN:978-3-319-43162-8. https://doi.org/10.1007/978-3-319-43162-8_5.

Devlin, J., Chang, M.-W., Lee, K., & Toutanova, K. (2019). Bert: Pre-training of deep bidirectional transformers for language understanding. In *NAACL-HLT*.

Di Gennaro, G., Buonanno, A., & Palmieri, F. A. (2021). Considerations about learning word2vec. *The Journal of Supercomputing, 77*(11), 12320–12335.

Ferjani, E., Hidri, A., Hidri, M. S., & Frihida, A. (2019). Mapreduce-based convolutional neural network for text categorization. In N. T. Nguyen, R. Chbeir, E. Exposito, P. Aniorté, & B. Trawiński (Eds.), *Computational collective intelligence* (pp. 155–166). Cham: Springer International Publishing. ISBN:978-3-030-28374-2.

Hadifar, A., & Momtazi, S. (2018). The impact of corpus domain on word representation: a study on persian word embeddings. *Language Resources and Evaluation, 52*(4), 997–1019.

Mikolov, T., Sutskever, I., Chen, K., Corrado, G., & Dean, J. (2013). Distributed representations of words and phrases and their compositionality. In *Proceedings of the 26th International Conference on Neural Information Processing Systems - Volume 2, NIPS'13*, USA (pp. 3111–3119). Curran Associates Inc.

Montenegro, J. L. Z., & da Costa, C. A. (2022). The hope model architecture: a novel approach to pregnancy information retrieval based on conversational agents. *Journal of Healthcare Informatics Research*, 1–42.

Pennington, J., Socher, R., & Manning, C. (2014). Glove: Global vectors for word representation. In *Proceedings of the 2014 Conference on Empirical Methods in Natural Language Processing (EMNLP)*, Doha, Qatar (pp. 1532–1543). Association for Computational Linguistics. https://doi.org/10.3115/v1/D14-1162.

Peters, M., Neumann, M., Iyyer, M., Gardner, M., Clark, C., Lee, K., & Zettlemoyer, L. (2018). Deep contextualized word representations. In *Proceedings of the 2018 Conference of the North American Chapter of the Association for Computational Linguistics: Human Language Technologies, Volume 1 (Long Papers)*, New Orleans, Louisiana (pp. 2227–2237). Association for Computational Linguistics. https://doi.org/10.18653/v1/N18-1202.

Vaswani, A., Shazeer, N., Parmar, N., Uszkoreit, J., Jones, L., Gomez, A. N., Kaiser, L. u., & Polosukhin, I. (2017). Attention is all you need. In I. Guyon, U. V. Luxburg, S. Bengio, H. Wallach, R. Fergus, S. Vishwanathan, & R. Garnett, (Eds.), *Advances in Neural Information Processing Systems* vol. 30 (pp. 5998–6008). Curran Associates, Inc.

Xie, H., Randall, M., & Chau, K.-w. (2022). Green roof hydrological modelling with gru and lstm networks. *Water Resources Management, 36*(3), 1107–1122.

Young, T., Hazarika, D., Poria, S., & Cambria, E. (2018). Recent trends in deep learning based natural language processing. *IEEE Computational IntelligenCe Magazine, 13*(3), 55–75.

Zhang, P., Xue, J., Lan, C., Zeng, W., Gao, Z., & Zheng, N. (2018). Adding attentiveness to the neurons in recurrent neural networks. In *Proceedings of the European Conference on Computer Vision (ECCV)* (pp. 135–151).

Zhu, Y., Kiros, R., Zemel, R., Salakhutdinov, R., Urtasun, R., Torralba, A., & Fidler, S. (2015). Aligning books and movies: Towards story-like visual explanations by watching movies and reading books. In *Proceedings of the IEEE International Conference on Computer Vision* (pp. 19–27).

Chapter 5
Question Answering over Text

Abstract This chapter provides a coherent and comprehensive overview of the available research studies in TextQA. Available studies are divided into two main categories: non-deep learning- based and deep learning-based approaches. A detailed explanation of the main property of each of these categories and available methods in each category will be covered in this chapter. Deep learning-based models will be studied in three different approaches, namely, interaction-based, representation-based, and hybrid models. Methods that belong to each of these three categories will be studied in detail. At the end of this chapter, a comparison of the proposed models is presented by discussing the strengths and weaknesses of these models. Also, we are going to explain the reasons behind the better performance of the state-of-the-art models. In the description of each research, the idea behind the model and its detailed architecture will be presented.

5.1 Introduction

TextQA or answer selection is defined as the task of selecting the relevant answer sentence among the available candidate answers. The goal of TextQA is retrieving the sentence which includes the answer to the given question without extracting the exact answer string. The available models solve this problem by calculating the probability of each candidate answer given the question. Then, the candidate answers are retrieved according to their probability.

Primary works in this area are non-deep learning approaches which rely on statistical information of the question and the answer sentences. These models try to capture the meaning of the question and the answer sentences by using lexical resources, such as WordNet, and creating language models. These models lack the ability to realize the contextual meaning of sentences which is crucial for finding the answer sentence. Besides that, they require external knowledge resources which are not easy to gather and keep up to date for all languages.

To mitigate this problem and create a semantically rich representation of the question and answer sentences based on their context, deep learning-based approaches are proposed. Deep learning-based models calculate the matching score

S. Momtazi, Z. Abbasiantaeb, *Question Answering over Text and Knowledge Base*,
https://doi.org/10.1007/978-3-031-16552-8_5

of the given question and candidate answer sentences. Some of the available models for QA TextQA use general text matching architectures, which can be divided into three categories based on their architecture: (1) representation-based models, (2) interaction-based models, and (3) hybrid models.

In representation-based models, semantic representation of each sentence is created separately. The semantic representations of sentences are compared to calculate the matching score or to classify the candidate answers as true or false. Interaction-based models calculate the matching score of the given question and candidate answer by modeling the interaction between their words. Most of them create an interaction matrix or tensor by comparing each pair of words from both sentences and use this matrix for measuring the matching score. Hybrid models are combinations of representation- and interaction-based models and contain both representation- and interaction-based components. Attention mechanism is considered as an interaction component and is mostly utilized in the hybrid models.

The remainder of this chapter is organized as follows: we will review the primary non-deep learning-based approaches proposed for the answer selection task in Sect. 5.2 and deep learning-based approaches in Sect. 5.3. The available deep learning-based approaches are studied in three distinct categories including representation-based, interaction-based, and hybrid models.

5.2 Non-deep Learning-Based Models

The translation model has been used in the retrieval models where we aim at finding the most relevant sentence to the given query. The translation model tries to extract the sentences which include both the translation of the query words and the exact query words. The idea of the translation model has been also used in QA. Murdock and Croft (2004) proposed a translation model for predicting the matching score between a question and an answer. The main intuition behind this model is that the exact question words may not appear in the answer sentence but they could be translated to some of the answer words. Murdock and Croft (2004) proposed the following language model for calculating the probability of question Q given the candidate answer A:

$$p(Q|A) = \prod_{i=1}^{m} \lambda(\sum_{j=1}^{n} p(q_i|a_j)p(a_j|A)) + (1 - \lambda)p(q_i|C) \tag{5.1}$$

where $p(q_i|a_j)$ denotes the probability of question term q_i translating to answer term a_j, $p(a_j|A)$ represents the probability of answer term a_j given the answer sentence A, λ is the smoothing parameter, $p(q_i|C)$ is a smoothing term which calculates the probability of question term q_i in corpus C, m is the question length, and n is the length of the answer sentence.

Summation of the translation probability of each term to all of its translations must be equal to one and is obtained by dividing one to count of its translations. The above formulation is mainly developed for the machine translation task where the exact matching among two different languages is not probable, while in the answer sentence ranking task, the chance of appearance of the exact question term in the answer sentence is high. Therefore, devoting less than one translation probability to the word, which is translated to itself, is not fair. To mitigate the mentioned problem, the following change is made:

$$p\left(q_i|a_j\right) p\left(a_j|A\right) = t_i\, p\left(q_i|A\right) + (1 - t_i) \sum_{1 \le j \le n, a_j \ne q_i} p\left(q_i|a_j\right) p\left(a_j|A\right) \tag{5.2}$$

where $t = 1$ means that the exact query term exists in the answer sentence.

Since answer sentences are not very long, it is very likely that part of the question terms does not appear in the answer sentence. In this case, the smoothing term from corpus collection is usually used. For better modeling the probability of unseen question terms, the document probability is also used for smoothing, which is defined as follows:

$$p(Q|D) = \prod_{i=1}^{m} \lambda p\left(q_i|D\right) + (1 - \lambda) p\left(q_i|C\right) \tag{5.3}$$

Finally, the document probability is merged with the answer probability for estimating the relevance score of the question and the answer pair as follows:

$$\begin{aligned} p(Q|A) = \prod_{i=1}^{m} \Big[& \beta(\lambda \sum_{j=1}^{n} p(q_i|a_j) p(a_j|A) + (1 - \lambda) p(q_i|C)) \\ & + (1 - \beta)(\lambda p(q_i|D_A) + (1 - \lambda) p(q_i|C)) \Big] \end{aligned} \tag{5.4}$$

where D_A represents the document containing the answer sentence A and C represents the collection which includes both documents and answer sentences.

In the word-based language model, exact term matching between the question and the answer sentences is considered, which results in missing relevant information that does not include question terms. Consider the following question and candidate answers:

Q: *Who invented the car?*

A_1: *Between 1832 and 1839, Robert Anderson of Scotland invented the first crude electric car carriage.*

A_2: *Nicolas-Joseph Cugnot built the first self-propelled mechanical vehicle.*
A_3: *An automobile powered by his own engine was built by Karl Benz in 1885 and granted a patent.*

The word-level unigram language model retrieves the first candidate answer with higher probability than the other candidate answers as it contains more question terms ("invented," "the," and "car"). This model lacks the ability of recognizing other relevant words in the second and the third answers including "vehicle," "automobile," and "built".

To mitigate the abovementioned problem, Momtazi and Klakow (2009) proposed the class-based language model which uses word clustering approach in language model-based retrieval. They clustered words using the Brown word clustering algorithm. Word clustering models capture various types of features (including syntactic, semantic, and statistical features) and cluster words based on them. In the class-based language model, probability of question given the answer is calculated based on the cluster information of the question terms as follows:

$$
\begin{aligned}
P(Q|S) &= \prod_{i=1}^{M} P(q_i|C_{q_i}, S)P(C_{q_i}|S) \\
P(C_{q_i}|S) &= \frac{f_S(C_{q_i})}{\sum_w f_S(w)}
\end{aligned}
\tag{5.5}
$$

where $P(C_{q_i}|S)$ is the probability of question term q_i's cluster given the answer sentence S and is calculated by dividing the count of answer terms belonging to this cluster over the count of all answer terms. To sum up, in the class-based language model, rather than mapping the question and answer terms, their clusters are matched.

The class-based model increases recall as it tries to retrieve all relevant sentences by considering word clusters. On the other hand, the precision may decrease due to extracting more irrelevant sentences. To solve this problem, they combined the class-based and the word-based language models as follows:

$$
P(Q|S) = \prod_{i=1...M} [\lambda P(q_i|C_{q_i}S)P(C_{q_i}|S) + (1-\lambda)P(q_i|S)] \tag{5.6}
$$

where λ is the interpolation parameter.

The unigram language models cannot capture the dependencies among words of the sentence as they are not sensitive to order of words. To overcome this problem, bigram and trigram class-based language models are proposed. The class-based n-grams are supposed to be more efficient than word-based n-grams, because observing a sequence of related words is more probable than observing the sequence of exact word n-grams in the answer sentence. Formulation of bigram and trigram class-based language models is as follows:

$$P^{bi}(Q|S) = P\left(q_1|C_{q_1}, S\right) P\left(C_{q_1}|S\right) \prod_{i=2...M} P\left(q_i|C_{q_i}, S\right) P\left(C_{q_i}|C_{q_{i-1}}, S\right) \tag{5.7}$$

$$\begin{aligned} P^{tri}(Q|S) =& P\left(q_1|C_{q_1}, S\right) P\left(C_{q_1}|S\right) P\left(q_2|C_{q_2}, S\right) P\left(C_{q_2}|C_{q_1}, S\right) \\ & \prod_{i=3...M} P\left(q_i|C_{q_i}, S\right) P\left(C_{q_i}|C_{q_{i-2}}C_{q_{i-1}}, S\right) \end{aligned} \tag{5.8}$$

In addition to the class-based language model, Momtazi and Klakow (2011) proposed the idea of a trained trigger language model for solving the problem of the word-level unigram language model. In the trained trigger language model, rather than considering the occurrence of question term in the answer sentence, the number of times that query term triggers each answer term is considered. The triggering probability of question term q_i given the answer term s_j is trained on corpus S and calculated as follows:

$$P_{\text{trigger}}(q_i|s_j) = \frac{f_C(q_i, s_j)}{\sum_{q_i} f_C(q_i, s_j)} \tag{5.9}$$

where $f_C(q_i, s_j)$ is the number of times the question term q_i triggers the answer term s_j in training corpus S. Having the above triggering probability, the probability of question term q_i given the answer sentence S is defined as follows:

$$P_{\text{trigger}}(q_i|S) = \frac{1}{N} \sum_{j=1}^{N} P_{\text{trigger}}(q_i|s_j) \tag{5.10}$$

and trained-trigger language model is defined as:

$$P_{\text{trigger}}(Q|S) = \left(\frac{1}{N}\right)^M \prod_{i=1}^{M} \sum_{j=1}^{N} \frac{f_C\left(q_i, s_j\right)}{\sum_q f_C\left(q_i, s_j\right)} \tag{5.11}$$

Five different types of triggering methods for calculating the triggering probability $P_{\text{trigger}}(q_i|s_j)$ are described in the following.

- **Inside sentence triggering:** The words which appear in the same sentence trigger each other. For training the inside sentence triggering model, an unannotated corpus is used. By utilizing this model, the answer sentences which contain words that appear in the same sentences as question terms are retrieved.
- **Across sentence triggering:** In this model, each word triggers all words in the next sentence in the training corpus. Using this triggering model, the answer sentences which contain words that appeared in the next sentence of the question terms in the training corpus are retrieved.

- **Question and answer pair triggering:** This model is trained on QA corpus. Each question term of the question sentence triggers all answer terms. Having the following example in the training corpus:

  Q: How high is Mount Hood?
  A: Mount Hood is in the Cascade Mountain range and is 11,245 feet.

  The model realizes that the word "high" triggers the word "feet" in the answer sentence. As a result, given the following question and answer pair:

  Q: How high is Everest?
  A: Everest is 29,029 feet.

  Even though the answer sentence shares just one similar word with the question sentence, the term "high" in the question sentence triggers the word "feet" in the answer sentence.
- **Self triggering (Momtazi & Klakow, 2015):** Each question word just triggers itself and, as a result, is similar to the word-level unigram language model.
- **Named entity triggering (Momtazi & Klakow, 2015):** In this approach, the question words trigger the named entity types of the answer terms. In other words, instead of triggering a specific word, the question terms trigger words with a specific named entity label. For training this model, a QA corpus with named entity labels is required. Consider the following question with candidate answers:

  Q: When was Einstein born?
  A_1: *Albert Einstein was born on 14 March 1879.*
  A_2: *Albert Einstein was born in Germany.*
  A_3: *Albert Einstein was born in a Jewish family.*

  All of the candidate answers are equally related to the question sentence according to the word-level language model as they include two main terms of the question sentence, namely, "Einstein" and "born." Using named entity triggering model, the first answer is retrieved with higher score as the word "when" triggers the words with "DATE" entity type and the first candidate answer includes "DATE" named entity type.

Similar to the class-based language model, it is expected that the trained trigger language model increases the system recall, as it captures a higher level of relationship among words. To avoid decreasing precision, the interpolation model is also defined for trained trigger language model as follows:

$$P(Q|S) = \prod_{i=1}^{M} [\lambda P_{\text{trigger}}(q_i|S) + (1-\lambda) P_{\text{word}}(q_i|S)] \tag{5.12}$$

where M is question length and λ is the weighting parameter.

5.3 Deep Learning-Based Models

5.3.1 *Representation-Based Models*

In the representation-based approach, different neural architectures have been used for creating the representation of the answer and the question sentences. The representation obtained by each neural architectures represents the input sentence by emphasizing on a specific view of the sentence. To be more specific, a representation can be influenced by all of the sentence words equally, can be influenced by the last words of a sentence, or can be influenced only by informative words. In addition, the generated representation can be based on local or long-term dependencies. In the following, we will discuss the architecture of several representation-based models and indicate which neural network they utilized for sentence representation.

Severyn and Moschitti (2015) proposed a model that uses CNN for modeling the semantic representation of the sentences. The special feature of this model is its ability in incorporating extra features. Each sentence is represented by using a single convolution layer and max-pooling. Representations of the query and the document sentences are concatenated with other auxiliary features, extracted from query and document sentences, and fed to a hidden layer. The given query and document sentence are classified using the softMax layer. Each layer of this model is explained below.

The word embeddings of the query and the document sentences are fed to CNN, and semantic representations of query (x_q) and document (x_d) sentences are built. Similarity of the query and the document sentences is computed according to the idea of noisy channel as follows:

$$\operatorname{sim}\left(x_q, x_d\right) = x_q^T M x_d \tag{5.13}$$

where $M \in R^{d \times d}$ represents the similarity matrix which transforms the document to the most near representation of the query sentence. Besides the x_{sim} feature, which captures the semantic and syntactic similarity of the given sentences, another feature named x_{feat} is also extracted. These two auxiliary features are concatenated to build the semantic representation of the query and document sentences ($x_{\text{join}} = \left[x_q^T; x_{\text{sim}}; x_d^T; x_{\text{feat}}^T\right]$) and passed to a fully connected layer followed by a softMax layer as follows:

$$\alpha\left(w_h \cdot x + b\right) \tag{5.14}$$

where w_h is the model parameter, b is the bias, and α represents the non-linear function which is substituted by softMax.

Yu et al. (2014) proposed a model that was one of the first attempts for solving the answer sentence selection task by using neural networks. Two distinct approaches are used for modeling the representation of sentences including bigram and bag-of-words. In the bag-of-words approach, CNN is used for modeling the sentence representation. In the model proposed by Yu et al. (2014), each question q_i is associated with a set of candidate answers $A = \{a_{i1}, a_{i2}, \cdots, a_{im}\}$ and labels $Y = \{y_{i1}, y_{i2}, \cdots, y_{im}\}$, indicating whether the candidate answer is true ($y = 1$) or not ($y = 0$). The obtained representations for the question (q) and the answer (a) sentences are matched for predicting the label of each question and answer pair as follows:

$$P(y = 1|\mathrm{q}, \mathrm{a}) = \sigma\left(\mathrm{q}_{\mathrm{m}}^{\mathrm{T}}\mathrm{Ma} + b\right) \tag{5.15}$$

where $q \in R^d$ and $a \in R^d$ are question and answer representations, respectively, $M \in R^{d \times d}$ is the model parameter, and b is the bias. This matching model works by creating a new representation for question sentence $q' = Ma$ based on the answer sentence and comparing it with the question's representation by dot product. This model is trained to learn matrix M by minimizing the cross-entropy loss of the predicted label and the actual label. In the following, two approaches proposed for creating the representation of each sentence are described.

- **Bag-of-words:** Representation of each sentence is generated by averaging the word embeddings as follows:

$$s = \sum_{i=1}^{|s|-1} \tanh\left(T_L s_i + T_R s_{i+1} + b\right) \tag{5.16}$$

where $|s|$ is the length of the given sentence and s_i represents the embedding of i-th term of the given sentence.

- **Bigram:** In bag-of-words model, sequence of words is not considered, and representation of sentence is generated without considering the dependency among words. Bigram model tries to capture local dependencies within a window of two consecutive words by using CNN. Each bigram is modeled into a new feature by CNN layer as follows:

$$\mathrm{c_i} = \tanh\left(\mathrm{T} \cdot \mathrm{s}_{\mathrm{i}:i+1} + \mathrm{b}\right) \tag{5.17}$$

where $T \in R^2$ is the convolution parameter, b is the bias, and c_i is the bigram feature. The above equation represents the calculations for each row of matrix T. The average of bigram features is calculated for creating the representation of the whole sentence by average pooling. Bigram model over all rows of matrix T followed by an average pooling is as follows:

$$s = \sum_{i=1}^{|s|-1} \tanh\left(T_L s_i + T_R s_{i+1} + b\right) \tag{5.18}$$

where $s \in R^d$ is the representation of a given sentence, $s_i \in R^d$ is the embedding of i-th term of the given sentence, $T_L \in R^{dtimesd}$ and $T_R \in R^{dtimesd}$ are model parameters, and b is the bias.

Holographic Dual LSTM (HD-LSTM) (Tay et al., 2017) models each of the question and answer sentences separately by LSTM. HD-LSTM follows holographic composition, which is parameterless and reduces time and computational complexity, for comparing the representation of question and answer embeddings. In HD-LSTM, representations of question and answer sentences are obtained from Q-LSTM and A-LSTM, respectively. Representations of question and answer sentences are compared in the holographic layer. Output of this layer is concatenated to overlap features and bilinear similarity of two sequences. This vector is passed through a holographic hidden layer followed by a fully connected layer to predict the matching score of question and answer sentences. HD-LSTM is trained according to point-wise approach by minimizing the cross-entropy loss function. Layers of HD-LSTM are explained in the following.

- **Learning QA representations:** Representation of question and answer sentences is modeled by two separate LSTMs, namely, Q-LSTM and A-LSTM. Each sentence is represented by a sequence of pre-trained Skip-gram embeddings of words. These embeddings are fed to LSTM, and the last hidden state of LSTM output is considered as the representation of the whole sentence.
- **Holographic matching of QA pairs:** A novel compositional model based on deep neural networks is proposed for modeling the relation among two given representations. Circular correlation of two vectors is used for modeling the relation of the two sentences as follows:

$$q \circ a = q \star a \tag{5.19}$$

where $\circ$ represents the compositional operators and $\star$ is the circular correlation. The circular correlation is defined as the summing product of elements in a specific circular path. Output of circular correlation has the same dimension of the input vectors. As the input vectors must have the same length, the shorter sequence is padded to the length of the longer sequence. Circulation correlation can be computed by each of the following equations:

$$[q \star a]_k = \sum_{i=0}^{d-1} q_i a_{(k+i) \bmod d} \quad q \star a = \mathcal{F}^{-1}(\overline{\mathcal{F}(q)} \odot \mathcal{F}(a)) \tag{5.20}$$

where q and a are representations of question and answer sentences, respectively, and $\mathcal{F}$ shows a fast Fourier transform.
- **Holographic hidden layer:** Result of holographic matching layer is augmented by using two additional features, namely, bilinear similarity and word overlap features. Bilinear similarity (*Sim*) is a scalar value and is computed by using a similarity matrix M. The bilinear similarity and word overlap features are

concatenated ($[q \star a]$) to form input of the holographic hidden layer as follows:

$$Sim(q, a) = \mathbf{q}^T M \mathbf{a} \quad (5.21)$$

$$h_{out} = \sigma\left(W_h \cdot \left[[q \star a], sim(q, a), X_{feat}\right] + b_h\right) \quad (5.22)$$

where W_h is the model parameter, X_{feat} is the word overlap feature, and b_h is the bias.

- **SoftMax layer:** Output of previous layer h_{out} is passed through a fully connected layer followed by a softMax layer for predicting the probability of answer given the question as follows:

$$\mathrm{P} = softMax\,(\mathrm{W_f.h_{out}} + \mathrm{b_f}) \quad (5.23)$$

where W_f and b_f are the model parameters.

Convolutional-pooling LSTM and **convolution-based LSTM** models (Tan et al., 2016) are based on the basic QA-LSTM model. These three models have representation-based architecture. In the basic QA-LSTM model, as is shown in Fig. 5.1, textual information of each sequence is extracted by LSTM. The main motivation behind proposing convolutional-pooling LSTM and convolution-based LSTM models is using both CNN and RNN for better modeling the contextual information of the given text. RNNs are designed to capture long-term dependencies by remembering past information, while CNNs are capable of extracting the local n-gram features. In convolutional-pooling LSTM, CNN is applied to the output hidden states of LSTM. Hidden states contain long-term dependencies within text, and CNN extracts local dependencies among the hidden states. In convolution-based LSTM, local n-grams or features are extracted from CNN, and they are passed to an RNN for extracting the long-term dependencies. Each of these models is explained briefly in the following.

- **Basic QA-LSTM:** Word embeddings of the question and the answer sentences are given to BiLSTM separately to generate hidden states. Two LSTMs are

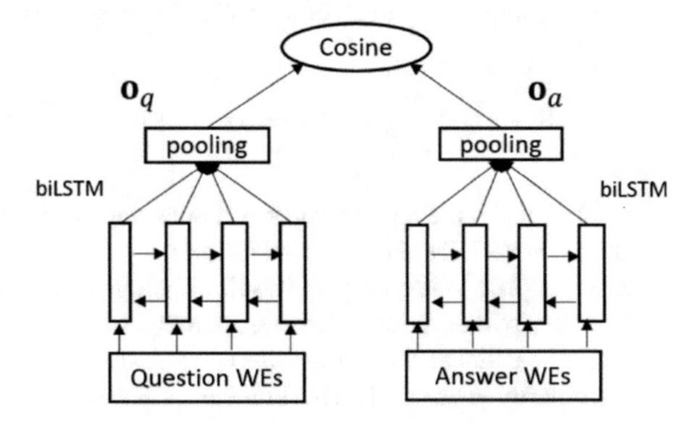

Fig. 5.1 The architecture of the basic QA-LSTM model (Tan et al., 2016) ("©1963–2022 ACL, reprinted with permission")

used for processing the given sequence in two forward and backward directions. Forward and backward hidden states are concatenated to form the representation of each token $h_t = \overrightarrow{h_t} \| \overleftarrow{h_t}$. Given a sequence of word embeddings $X = \{\mathbf{x}(1), \mathbf{x}(2), \cdots, \mathbf{x}(n)\}$ as input, output of LSTM at each time step h_t is computed as follows:

$$\begin{aligned}
i_t &= \sigma\left(\mathbf{W}_i \mathbf{x}(t) + \mathbf{U}_i \mathbf{h}(t-1) + \mathbf{b}_i\right) \\
f_t &= \sigma\left(\mathbf{W}_f \mathbf{x}(t) + \mathbf{U}_f \mathbf{h}(t-1) + \mathbf{b}_f\right) \\
o_t &= \sigma\left(\mathbf{W}_o \mathbf{x}(t) + \mathbf{U}_o \mathbf{h}(t-1) + \mathbf{b}_o\right) \\
\tilde{C}_t &= \tanh\left(\mathbf{W}_c \mathbf{x}(t) + \mathbf{U}_c \mathbf{h}(t-1) + \mathbf{b}_c\right) \\
C_t &= i_t * \tilde{C}_t + f_t * C_{t-1} \\
\mathbf{h}_t &= o_t * \tanh\left(C_t\right)
\end{aligned} \tag{5.24}$$

where i, f, and o are the input, forget, and output gates; $\mathbf{W} \in R^{H \times E}$, $\mathbf{U} \in R^{H \times H}$, and $\mathbf{b} \in R^{H \times 1}$ are the model parameters; σ is the sigmoid function; and C_t is the cell memory network.

The output hidden states are used in three ways to produce the fixed-size vector representation of the given sentences: (1) concatenating last hidden states in both directions, (2) average pooling of all hidden states of BiLSTM, and (3) max-pooling over all output vectors. The fixed-size representations of question and answer sentences (q and a) are compared by cosine similarity to compute the matching score of question and answer sentences.

- **Convolutional-pooling LSTM:** As the architecture of convolutional-pooling LSTM is shown in Fig. 5.2, CNN is used instead of the pooling layer (in QA-LSTM model) over output hidden states of BiLSTM. Input of CNN is matrix $Z \in \mathcal{R}^{k|h| \times L}$ which is constructed by concatenating k hidden states produced by BiLSTM. Column m in matrix Z is the concatenation of k hidden states with size $|h|$ centralized in the m-th token of sequence. Output of CNN layer, which

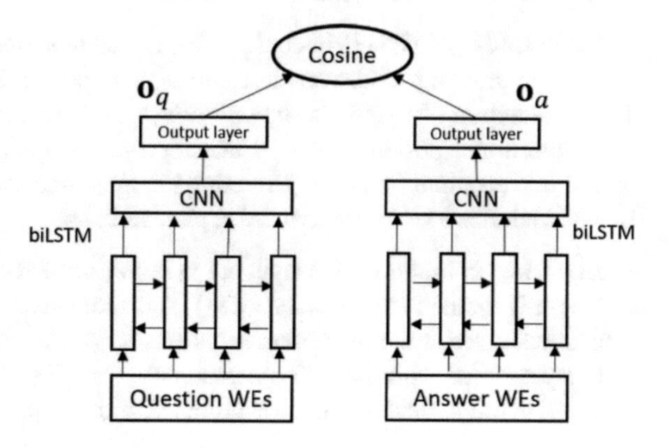

Fig. 5.2 The architecture of the convolutional-pooling LSTM model (Tan et al., 2016) ("©1963–2022 ACL, reprinted with permission")

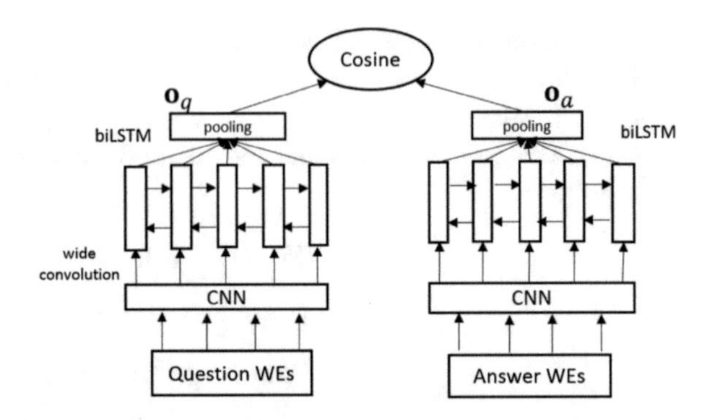

Fig. 5.3 The architecture of the convolution-based LSTM model (Tan et al., 2016) ("©1963–2022 ACL, reprinted with permission")

employs c filters, is used for generating representation vectors of question and answer sentences o_q and o_a, respectively, as follows:

$$C = \tanh\left(\mathbf{W}_{cp} Z\right) \tag{5.25}$$

$$\left[o_j\right] = \max_{1<l<L}\left[C_{j,l}\right] \tag{5.26}$$

where W_{cp} is the model parameter and $C \in \mathcal{R}^{c \times L}$ is the output of convolution with c filters.

- **Convolution-based LSTM:** According to Fig. 5.3, local n-gram features are extracted by using CNN over input sequence. Then, local n-grams are passed to BiLSTM for extracting long-term dependencies. Input of the CNN layer, matrix $D \in \mathcal{R}^{kE \times L}$, is built by using word embeddings. Column l in this matrix is built by concatenating embeddings of k words centralized in l-th word. Embedding of each word has size E. Output of convolution ($\mathbf{X} \in \mathcal{R}^{c \times L}$) is generated as follows:

$$\mathbf{X} = \tanh\left(\mathbf{W}_{cb} D\right) \tag{5.27}$$

  The matrix X is passed to BiLSTM. Max-pooling is used over output hidden states of BiLSTM to generate the representation of each sentence.

Basic CNN (BCNN) (Yin et al., 2016) model has Siamese architecture and uses CNN for extracting robust features and local dependencies in the given sequence. The architecture of BCNN is shown in Fig. 5.4. Each sentence is modeled through convolution and pooling layers. Final representations of the given sentences are passed to the output layer for predicting the matching score of the given sentences. BCNN includes four layers that are explained below.

- **Input layer:** Inputs of this layer are s_0 and s_1 sentences. The maximum sequence length is defined as $s = \max(s_0, s_1)$, and sequences shorter than this length are padded to reach the maximum sequence length. Each word within the sentence is represented with the 300-dimensional ($d = 300$) Word2Vec (Mikolov et al., 2013) vectors. Each sentence is modeled with a representation map $\in R^{d_0 \times s}$.

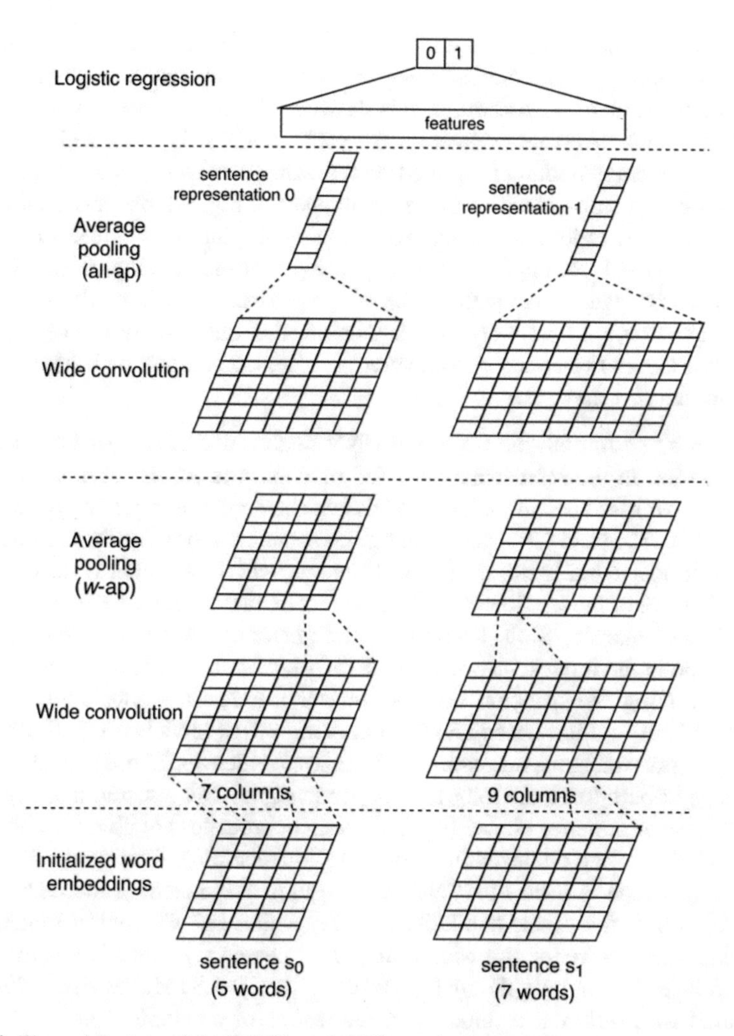

Fig. 5.4 The architecture of the BCNN model (Yin et al., 2016) ("©1963–2022 ACL, reprinted with permission")

- **Convolution layer:** Each sentence contains a set of words $v_1, v_2, \cdots, v_s$, and convolution is applied on a window of w consecutive words. Window of w words, or phrase with length w, is represented as $\mathbf{c}_i \in \mathbb{R}^{w \cdot d_0}$. Representation of each phrase $\mathbf{c}_i$ denoted by $p_i \in R^{d_1}$ is generated by applying the convolution weight $W \in R^{d_1 \times w.d_0}$ as follows:

$$P_i = tanh(W.c_i + b) \tag{5.28}$$

where each window contains words $v_{i-w+1}, \cdots, v_i$ where embedding of $v_j,\ j < 0\ or\ j > s$ is set to zero, and $b \in R^{d_1}$ is the bias.

- **Average pooling layer:** Two pooling mechanisms, namely, $w - ap$ and $all - ap$ average pooling, are applied after the middle and last convolution layers, respectively. In $all - ap$ pooling, as is shown in Fig. 5.4, by column-wise average over all columns, a representation for each sentence is generated. In $w - ap$ pooling, average pooling is applied on the window of w columns, and a feature map is generated. According to the second part of Fig. 5.4, the convolution layer transforms the input feature map with size $d_0 \times s$ into a feature map with size $d_0 \times s + w - 1$. By applying $w - ap$ pooling over the output of the convolution layer, a new feature map with size $d_0 \times s$ is generated. Each middle convolution layer captures a different level of abstraction from the given sequence.
- **Output layer:** For binary classification task, logistic regression is applied to the output of $all - ap$ layer.

As can be seen, both the CNN and RNN neural models are used for sentence representation in the representation-based models. The BCNN (Yin et al., 2016) and bag-of-words (Yu et al., 2014) models and the model proposed by Severyn and Moschitti (2015) used CNN for creating the sentence representation. Using CNN helps the model to better extract the local dependencies in the input sentence. Due to the limited size of the window in CNN, it cannot extract the long-term dependencies in the input sentence, while RNNs have the power of memorizing the long-term information in the input sentence. HD-LSTM (Tay et al., 2017) model uses LSTM for representing the question and the answer sentences. Convolutional-pooling LSTM and convolution-based LSTM (Tan et al., 2016) models use both RNN and CNN for representing the question and the answer sentences. These two models take advantage of both local and long-term dependencies. After generating the question and the answer sentence embeddings, different approaches are used for comparing them and calculating the matching score. Machine learning algorithms like logistic regression, which is used in BCNN, can be used for matching the embeddings. QA-LSTM uses the cosine similarity of the question and the answer sentences as matching score. Severyn and Moschitti (2015) passed the embeddings to a feed-forward layer to compute the matching score. In HD-LSTM, the embeddings are compared with different methods, and the results of comparison are passed to a feed-forward layer for computing the matching score.

5.3.2 *Interaction-Based Models*

The interaction-based models calculate the matching score by interacting the question and the answer sentences. The interaction components use various approaches for comparing the question and the answer terms. We will discuss the architecture and the novelty of some interaction-based models in the following.

Stacked BiLSTM Wang and Nyberg (2015) was one of the earliest attempts in using neural networks for answer selection. The barrier of using external knowledge resources like WordNet and feature extraction was removed. Lexical matching is not enough for answer selection task, and extracting the contextual meaning of the words and sentences is required for answering the questions. For example, consider the following question and candidate answers:

Q: *What sport does Jennifer Capriati play?*

A_1: *Capriati, 19, who has not played competitive tennis since November 1994, has been given a wild card to take part in the Paris tournament which starts on February 13.*

A_2: *Capriati also was playing in the U.S. Open semifinals in '91, one year before Davenport won the junior title on those same courts.*

Although both answers include "Capriati" and "play" words and match the question lexically, just the first answer is correct because its meaning matches the question. Recurrent neural networks can remember information of the previous inputs in a sequence and use it in the current input. By using RNNs, a representation based on previous tokens of the sequence is generated for each input token. In this way, the sequence is just processed in one direction, while in a sentence, words can also influence the meaning of each other in the backward direction. To overcome this issue, two distinct RNNs can be used in the forward and backward directions, and by concatenating the hidden states of both directions, a more meaningful representation of each token is available. The information received from the previous input and passed to the next input is controlled by input and output gates. In stacked LSTM, the output sequence of one LSTM is passed as input to another LSTM which helps to extract higher levels of abstraction in the given sequence. Wang and Nyberg (2015) used stacked bidirectional LSTM (BiLSTM) for modeling contextualized meaning of the question and answer sentences. A version of LSTM without peephole connections is used with the following implementation:

$$\begin{aligned}
i_t &= \sigma\left(W_{xi}x_t + W_{hi}h_{t-1} + b_i\right)\\
f_t &= \sigma\left(W_{xf}x_t + W_{hf}h_{t-1} + b_f\right)\\
c_t &= f_t c_{t-1} + i_t \tau\left(W_{xc}x_t + W_{hc}h_{t-1} + b_c\right)\\
o_t &= \sigma\left(W_{xo}x_t + W_{ho}h_{t-1} + b_o\right)\\
h_t &= o_t \theta\left(c_t\right)
\end{aligned} \tag{5.29}$$

where i, f, and o are input, forget, and output gates, c is a cell memory activation vector, and τ and θ are activation functions. In this model, the *tanh* activation function is used. Final representation of each token y_t is obtained by using both forward and backward representations as follows:

$$y_t = W_{\overrightarrow{h}y}\overrightarrow{h_t} + W_{\overleftarrow{h}y}\overleftarrow{h_t} + b_y \tag{5.30}$$

where $\overrightarrow{h_t}$ and $\overleftarrow{h_t}$ are forward and backward hidden states at step t, respectively, $W_{\overrightarrow{h}y}$ and $W_{\overleftarrow{h}y}$ are the model parameters, and b_y is the bias.

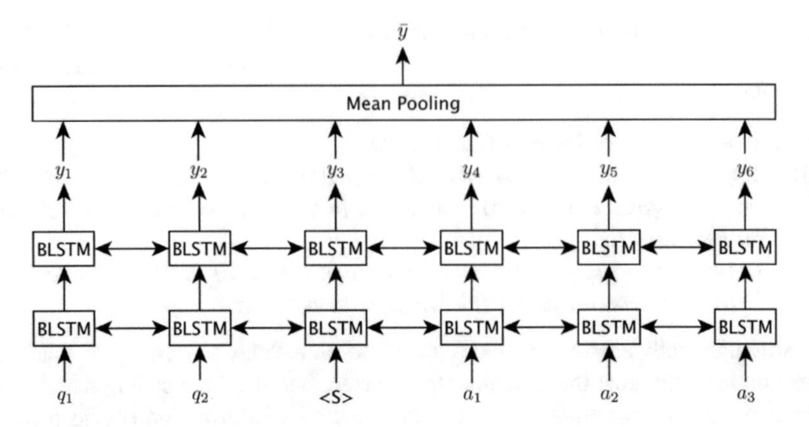

Fig. 5.5 The architecture of the stacked BiLSTM model (Wang & Nyberg, 2015) ("©1963–2022 ACL, reprinted with permission")

As the architecture of their model is shown in Fig. 5.5, the embeddings of the question and the answer sentences are connected with a $< S >$ token and processed by stacked BiLSTM. Mean-pooling over corresponding output to each hidden state y_t is considered as a matching score. In the test set, differentiation pooling strategies are used and compared.

The representation generated by distributed representation models, like Word2Vec, maps proper nouns with the same type and cardinal numbers to near embeddings and cannot distinguish them well. Sometimes, absence of a proper noun in the candidate answer is enough reason for discarding that answer. In this case, distributed representations mislead the model and may suggest a sentence with another proper noun with the same type which is not informative. For example, consider a question asking about *Japan* country; then, we expect the word *Japan* to occur in the answer sentence. Appearance of the word *China* in the candidate answer not only does not distinguish the difference between two sentences but also maps them to a near space. Output of stacked BiLSTM is enhanced by incorporating the matching between cardinal numbers and proper nouns by using gradient boosted regression tree (GBDT).

Pair-Wise Word Interaction Modeling (PWIM) (He & Lin, 2016) uses word interactions for measuring the question and the answer similarities. In PWIM, a novel similarity focus layer and a pair-wise word interaction modeling layer are proposed for measuring the similarity of two given sentences. As the architecture of PWIM is shown in Fig. 5.6, it contains four main layers. PWIM models each word regarding its context by using a BiLSTM and then calculates the word interactions according to their contextual representation. Output of word interaction modeling layer is a cube which is passed to similarity focus layer followed by a similarity classification layer for measuring the similarity of two given sentences. Each of these layers is described below.

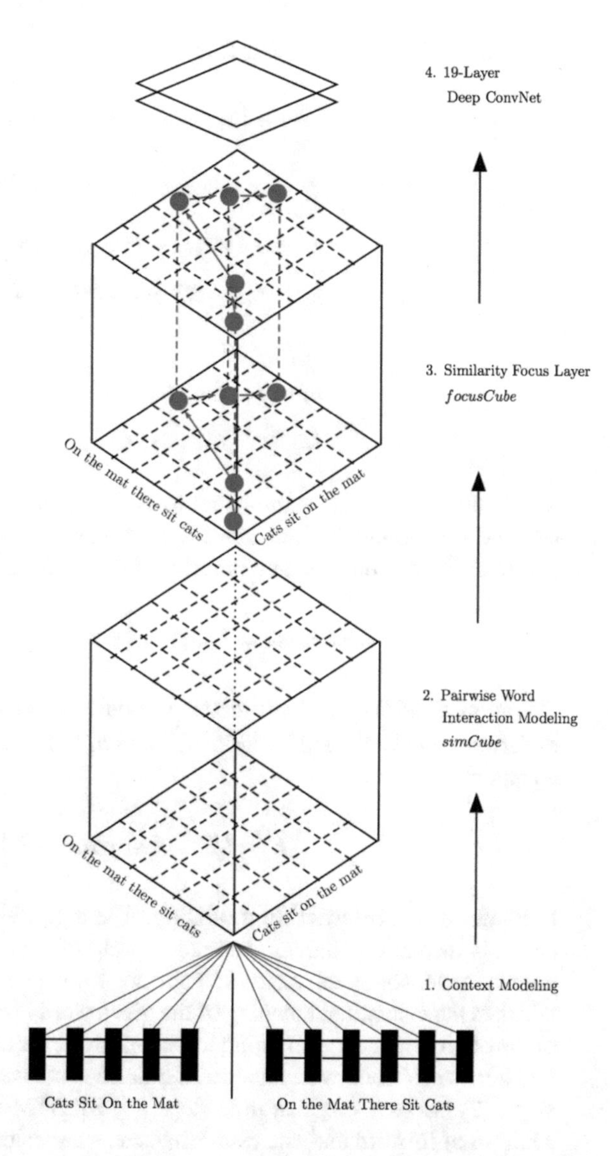

Fig. 5.6 The architecture of the PWIM model (He & Lin, 2016) ("©1963–2022 ACL, reprinted with permission")

- **Context modeling:** Words with similar context are more probable to be similar, and due to this fact, contextual meanings of words are created by using BiLSTM. BiLSTM includes two distinct LSTMs which process the given sentence in forward and backward directions. LSTM controls the flow of information through input (i), forget (f), and output (o) gates. In LSTM, given the input at time step t and output hidden state of the previous time step ($h_(t-1)$), the hidden state at time step t (h_t) is calculated as follows:

$$\begin{aligned}
i_t &= \sigma\left(W^i x_t + U^i h_{t-1} + b^i\right)\\
f_t &= \sigma\left(W^f x_t + U^f h_{t-1} + b^f\right)\\
o_t &= \sigma\left(W^o x_t + U^o h_{t-1} + b^o\right)\\
u_t &= \tanh\left(W^u x_t + U^u h_{t-1} + b^u\right)\\
c_t &= i_t \cdot u_t + f_t \cdot c_{t-1}\\
h_t &= o_t \cdot \tanh\left(c_t\right)\\
LSTM\left(x_t, h_{t-1}\right) &= h_t
\end{aligned} \tag{5.31}$$

where W^* and U^* are parameters to be learned, b^* is the bias, and σ is the sigmoid activation function. Outputs of forward and backward LSTMs ($LSTM^f$ and $LSTM^b$) are concatenated to form the output for BiLSTM model as follows:

$$BiLSTMs\left(x_t, h_{t-1}\right) = \left\{LSTM^f, LSTM^b\right\} \tag{5.32}$$

Output of BiLSTM at each time step t could be separated to obtain the output of forward and backward LSTMs (h_t^{for} and h_t^{back}) at time step t by the following equation:

$$h_t^{for}, h_t^{back} = unpack\left(h_t^{bi}\right) \tag{5.33}$$

- **Pair-wise word interaction modeling:** The algorithm of pair-wise word interaction is inspired by the human reader's way of comparing two sentences. As a human reader compares words and phrases among two sentences, each time step which is the contextual meaning of the given word is compared against all words of the other sentence. According to Algorithm 1, for calculating the similarity of two time steps from two sentences, similarities between (1) forward hidden time steps, (2) backward hidden time steps, (3) BiLSTM hidden time steps, and (4) addition of forward and backward time steps are calculated. For comparing two given hidden states $\overrightarrow{h_1}$ and $\overrightarrow{h_2}$, three different similarity measurement metrics are used as follows:

$$coU(\overrightarrow{h_1}, \overrightarrow{h_2}) = \{\cos(\overrightarrow{h_1}, \overrightarrow{h_2}), L_2 Euclid(\overrightarrow{h_1}, \overrightarrow{h_2}), DotProduct(\overrightarrow{h_1}, \overrightarrow{h_2})\} \tag{5.34}$$

Cosine similarity metric (cos) measures the angle between two vectors; $Euclid$ and $DotProduct$ metrics measure the magnitude difference of two vectors. The output of Algorithm 1 is a cube with size $R^{13.|sent_1|.|sent_2|}$. For each pair of words from both sentences, 12 similarity measures are calculated, and 1 extra padding is considered.

Algorithm 1 Pair-wise word interaction modeling (He & Lin, 2016)

Initialize: $SimCube \in R^{13.|sent_1|.|sent_2|}$ to all 1
for *each time step* $t = 1...|sent_1|$ **do**
  for *each time step* $s = 1...|sent_2|$ **do**
    $h_{1t}^{for}, h_{1t}^{back} = unpack\left(h_{1t}^{bi}\right)$
    $h_{2s}^{for}, h_{2s}^{back} = unpack\left(h_{2s}^{bi}\right)$
    $h_{1t}^{add} = h_{1t}^{for} + h_{1t}^{back}$
    $h_{2s}^{add} = h_{2s}^{for} + h_{2s}^{back}$
    $simCube[1:3][t][s] = coU\left(h_{1t}^{bi}, h_{2s}^{bi}\right)$
    $simCube[4:6][t][s] = coU\left(h_{1t}^{for}, h_{2s}^{for}\right)$
    $simCube[7:9][t][s] = coU\left(h_{1t}^{back}, h_{2s}^{back}\right)$
    $simCube[10:12][t][s] = coU\left(h_{1t}^{add}, h_{2s}^{add}\right)$
  end for
end for

- **Similarity focus layer:** The word interactions, calculated in the previous layer, are passed to similarity focus layer. According to Algorithm 2, important word interactions are recognized and re-weights by giving higher weight. The weight of each interaction is calculated and stored in *mask* cube according to Algorithm 2. The most important word interactions are given weight 1, and other non-important interactions are given weight 0.1, while no words are recognized as important more than one time in both sentences. *calcPos* functions return the position of the given interaction in *simCube*.
- **Similarity classification:** The *simCube* obtained in the previous layer is considered as an image. A 19-layer CNN, consisting of spatial convolution and pooling layers, is used for modeling the similarity of two sentences based on the interactions.

Match-SRNN (Wan et al., 2016b) uses a recursive matching structure for modeling the interaction among two given sentences. In a recursive matching algorithm, different combinations of available prefixes are compared and used. For modeling the interaction among two $S_1[1:i] = \{w_1, \cdots, w_i\}$ and $S_2[1:j] = \{v_1, \cdots, v_j\}$ prefixes from S_1 and S_2 sentences, three other phrase-level interactions $\overrightarrow{h}_{i-1,j}$, $\overrightarrow{h}_{i,j-1}$, and $\overrightarrow{h}_{i-1,j-1}$ with a word-level interaction $\overrightarrow{s}(w_i, v_j)$ are used as follows:

$$\overrightarrow{h}_{ij} = f(\overrightarrow{h}_{i-1,j}, \overrightarrow{h}_{i,j-1}, \overrightarrow{h}_{i-1,j-1}, \overrightarrow{s}(w_i, v_j)) \tag{5.35}$$

To better comprehend the performance of a recursive matching algorithm, consider two sentences 1) "The cat sat on the mat" and 2) "The dog played balls

Algorithm 2 Forward pass: similarity focus layer (He & Lin, 2016)

Input: $simCube \in R^{13 \cdot |sent_1| \cdot |sent_2|}$
Initialize: $mask \in R^{13 \cdot |sent_1| \cdot |sent_2|}$ to all 0.1
Initialize: $s1tag \in R^{|sent_1|}$ to all zeros
Initialize: $s2tag \in R^{|sent_2|}$ to all zeros
$sortIndex_1 = sort(simCube[10])$
for $each\ id = 1 \ldots |sent_1| + |sent_2|$ **do**
  $pos_{s1}, pos_{s2} = calcPos\,(id, sortIndex_1)$
  if $s1tag\left[pos_{s1}\right] + s2tag\left[pos_{s2}\right] == 0$ **then**
    $s1tag\left[pos_{s1}\right] = 1$
    $s2tag\left[pos_{s2}\right] = 1$
    $mask[:]\left[pos_{s1}\right]\left[pos_{s2}\right] = 1$
  end if
end for
Re-Initialize: $s1tag$, $s2tag$ to all zeros
$sortIndex_2 = sort(simCube[11])$
for $each\ id = 1 \ldots |sent_1| + |sent_2|$ **do**
  $pos_{s1}, pos_{s2} = calcPos\,(id, sortIndex_2)$
  if $s1tag\left[pos_{s1}\right] + s2tag\left[pos_{s2}\right] == 0$ **then**
    $s1tag\left[pos_{s1}\right] = 1$
    $s2tag\left[pos_{s2}\right] = 1$
    $mask[:]\left[pos_{s1}\right]\left[pos_{s2}\right] = 1$
  end if
end for
$mask[13][:][:] = 1$
$focusCube = mask \cdot simCube$
Return: $focusCube \in R^{13 \cdot |sent_1| \cdot |sent_2|}$

on the floor." For comparing two prefixes $S_1[1:3]$: *The cat sat* and $S_2[1:4]$: *The dog played balls*, the interactions among the following prefixes are used:

- $S_1[1:2]$: *The cat* and $S_2[1:4]$: *The dog played balls*,
- $S_1[1:3]$: *The cat sat* and $S_2[1:3]$: *The dog played*,
- $S_1[1:2]$: *The cat* and $S_2[1:3]$: *The dog played*

which are defined as $\overrightarrow{h}_{2,4}$, $\overrightarrow{h}_{3,3}$, and $\overrightarrow{h}_{2,3}$, respectively. Also, the interaction among two words *sat* and *balls* is used as $\overrightarrow{s}(sat, balls)$ in the above formulation.

According to the architecture of the Match-SRNN model, word-level interaction among two given sentences is modeled by using a neural tensor network, and the recursive matching strategy is applied by using a spatial RNN model. Output of the spatial RNN layer is passed to a linear scoring function for modeling the semantic matching score of the given sentences. Each of these layers is described below.

- **Neural tensor network:** The word-level interactions are captured by using a neural tensor network which obtains more complicated interactions. Interaction among two words w_i and v_j with their distributed representations $u(w_i)$ and

$u(v_j)$ can be computed as follows:

$$\overrightarrow{s}_{ij} = F\left(u(w_i)^T T^{[1:c]} u(v_j) + W\begin{bmatrix} u(w_i) \\ u(v_j) \end{bmatrix} + \overrightarrow{b}\right) \tag{5.36}$$

where $T^i, i \in [1, \cdots, c]$ represents one slice of tensor parameters T, W is the model parameter, and $\overrightarrow{b}$ is the bias.

- **Spatial RNN:** Spatial RNN is a multidimensional RNN which is suitable for modeling the recursive matching. Among various available RNNs including basic RNN, GRU, and LSTM, GRU is used due to its easy implementation. The spatial-GRU has four updating gates and three reset gates for capturing interaction of three other prefixes, namely, $\overrightarrow{h}_{i-1,j}$, $\overrightarrow{h}_{i,j-1}$, and $\overrightarrow{h}_{i-1,j-1}$. Spatial-GRU models the function f in recursive matching strategy, which computes the interaction $\overrightarrow{h}_{i,j}$, as follows:

$$\begin{aligned}
&\mathbf{q}^T = \left[\mathbf{h}_{i-1,j}^T, \mathbf{h}_{i,j-1}^T, \mathbf{h}_{i-1,j-1}^T, \mathbf{s}_{ij}^T\right]^T \\
&\mathbf{r}_l = \sigma\left(W^{(r_l)}\mathbf{q} + \mathbf{b}^{(r_l)}\right), \quad \overrightarrow{r_t} = \sigma\left(W^{(r_t)}\mathbf{q} + \mathbf{b}^{(r_t)}\right) \\
&\mathbf{r}_d = \sigma\left(W^{(r_d)}\mathbf{q} + \mathbf{b}^{(r_d)}\right), \quad \mathbf{r}^T = \left[\mathbf{r}_l^T, \mathbf{r}_t^T, \mathbf{r}_d^T\right]^T \\
&\mathbf{z}_i' = W^{(z_i)}\mathbf{q} + \mathbf{b}^{(z_i)}, \quad \overrightarrow{z_l'} = W^{(z_l)}\mathbf{q} + \mathbf{b}^{(z_l)} \\
&\mathbf{z}_t' = W^{(z_t)}\mathbf{q} + \mathbf{b}^{(z_t)}, \quad \mathbf{z}_d' = W^{(z_d)}\mathbf{q} + \mathbf{b}^{(z_d)} \\
&[\mathbf{z}_i, \mathbf{z}_l, \mathbf{z}_t, \mathbf{z}_d] = \text{softMaxByRow}\left(\left[\mathbf{z}_i', \mathbf{z}_l', \mathbf{z}_t', \mathbf{z}_d'\right]\right) \\
&\mathbf{h}_{ij}' = \phi\left(W\mathbf{s}_{ij} + U\left(\mathbf{r} \odot \left[\mathbf{h}_{i,j-1}^T, \mathbf{h}_{i-1,j}^T, \mathbf{h}_{i-1,j-1}^T\right]^T\right) + \mathbf{b}\right) \\
&\mathbf{h}_{ij} = \mathbf{z}_l \odot \mathbf{h}_{i,j-1} + \mathbf{z}_t \odot \mathbf{h}_{i-1,j} + \mathbf{z}_d \odot \mathbf{h}_{i-1,j-1} + \mathbf{z}_i \odot \mathbf{h}_{ij}'
\end{aligned} \tag{5.37}$$

where $\mathbf{z}_l$, $\mathbf{z}_t$, $\mathbf{z}_d$, and $\mathbf{z}_i$ are update gates; $\mathbf{r}_l$, $\mathbf{r}_t$, and $\mathbf{r}_d$ are reset gates; $W^{(x_y)}$s are model parameters; and $b^{(x_y)}$s are the biases.

- **Linear scoring function:** The last cell in the output of spatial-GRU ($\mathbf{h}_{mn}$) is considered as the complete interaction of two given sentences as it contains the information of interaction between whole prefixes. $\mathbf{h}_{mn}$ is passed through feed-forward neural network for modeling the interaction among two sentences as follows:

$$M(S_1, S_2) = W^{(s)} h_{mn} + b^{(s)} \tag{5.38}$$

where $W^{(s)}$ is the model parameter and $b^{(s)}$ is the bias.

MV-LSTM Wan et al. (2016a) considers the long-term and short-term dependencies in representing the tokens for interacting with them. Word-level, phrase-level, and sentence-level granularity representations do not carry the context information of terms efficiently, and they can lead to incorrect results. For example, consider the question "Which teams won top three in the World Cup?" and the following candidate answers:

A_1: *Germany is the champion of the World Cup.*
A_2: *The top three of the European Cup are Spain, the Netherlands, and Germany.*
A_3: *The top three attendees of the European Cup are from Germany, France, and Spain.*

In answering the mentioned question, "world cup" and "top three" words play an important role, and one of them has appeared in each of the A_1 and A_2 candidate answers. Therefore, the word-level representation of the first two sentences cannot help the model well to distinguish the correct answer A_2 from the incorrect answer A_1. On the other hand, by considering the phrase-level representations, the difference in contextualized meaning of the phrase "top three" in the two candidate answer sentences won't be obvious. For mitigating the mentioned problems, MV-LSTM utilizes BiLSTM for modeling the representation of the given sentence in each position regarding the whole sentence. LSTM is the best choice for this task, as it considers the long-term and short-term dependencies in generating the representation of each token within the sentence. According to the architecture of MV-LSTM, after generating the representation of each token within each sequence, they are compared using different comparison functions, and the obtained results are aggregated for calculating the matching score of given question and answer pairs. MV-LSTM includes three main layers which are explained in the following.

- **Positional sentence representation:** Two distinct LSTMs are used for processing the text in two forward and backward directions. Each of LSTMs generates the representation of the sequence in that direction by emphasizing on each word where influence of the adjacent terms depends on their distance. Given the sentence $S = (x_0, x_1, \cdots, x_T)$, two distinct representations of the sentence in each token are generated by the forward and backward LSTMs as $\mathbf{h}_t$ and $\overleftarrow{h_t}$. A special type of LSTM without peephole connections is used. The formulation for generating each h_t by an LSTM is as follows:

$$
\begin{aligned}
i_t &= \sigma\left(W_{xi}x_t + W_{hi}h_{t-1} + b_i\right) \\
f_t &= \sigma\left(W_{xf}x_t + W_{hf}h_{t-1} + b_f\right) \\
c_t &= f_t c_{t-1} + i_t \tanh\left(W_{xc}x_t + W_{hc}h_{t-1} + b_c\right) \\
o_t &= \sigma\left(W_{xo}x_t + W_{ho}h_{t-1} + b_o\right) \\
h_t &= o_t \tanh\left(c_t\right)
\end{aligned}
\tag{5.39}
$$

where f, i, and o are forget, input, and output gates, respectively. c is the stored information, and h is generated representation. Each token is represented by concatenating the forward and backward representations as $p_t = \left[\overrightarrow{h_t}^T, \overleftarrow{h_t}^T\right]^T$.

- **Interactions between two sentences:** Interaction tensor is built by comparing each token from the first sequence P_{X_i} with each token of the other sequence P_{Y_j}. Three different similarity functions, namely, cosine, bilinear, and tensor function, are used for comparing the tokens and building the interaction tensor. Bilinear and tensor similarity functions are as follows:

$$S(u, v) = u^t M v + b \tag{5.40}$$

$$s(u, v) = f\left(u^T M^{[1:c]} v + W_{uv} \begin{bmatrix} u \\ v \end{bmatrix} + b\right) \tag{5.41}$$

where M^i indicates a slice of the tensor parameter, f is the rectifier function, W_{uv} is the parameter, and b is the bias. Type of the output of the bilinear and cosine functions is matrix, whereas output of the tensor function is a tensor.

- **Interaction aggregation:** In the interaction matrices, top k values of the interaction matrix are concatenated to form the vector q. In the interaction tensor, top k values of each slice of tensor are concatenated to form the vector q. Finally, the obtained vectors from interaction matrices and interaction tensor are concatenated to form the feature vector q. The feature vector q is passed to a multi-layer perceptron (MLP) to calculate the matching score r as follows:

$$r = f(W_r q + b_r), s = W_s r + b_s \tag{5.42}$$

where W_s and W_r are the learnable parameters and b_r and b_s are the biases.

Gated Self-Attention Memory Network (GSAMN) (Lai et al., 2019) follows a different approach from the well-known compare-aggregate architecture, which is frequently used in answer sentence selection task, by proposing a memory-based gated self-attention mechanism for representing the question and the answer sentences. The representations of the question and the answer sentences are created by using GSAMN. In gated self-attention mechanism (GSAM), despite the usual attention mechanism, which calculates attention weights by normalized dot product of the terms from two sequences, association of two input vectors is calculated by gate vector according to:

$$\mathbf{g}_i = \sigma\left(f\left(\mathbf{c}, \mathbf{x}_i\right)\right) \tag{5.43}$$

where c is the context vector, $X = [\mathbf{x}_1..\mathbf{x}_n]$ is the input vector, σ is the sigmoid function, and g_i is the gate vector. The function f in GSAM is designed to incorporate all of the sequence terms rather than just one term with the context vector. In other words, the value of the gate vector relies on the entire input sequence and the context vector. The gate vector g_i for input x_i is calculated as follows:

$$
\begin{aligned}
\mathbf{v}^j &= \mathbf{W}\mathbf{x}_j + \mathbf{b};\ \mathbf{v}^c = \mathbf{W}\mathbf{c} + \mathbf{b} \\
s_i^j &= \mathbf{x}_i^T \mathbf{v}^j;\ s_i^c = \mathbf{x}_i^T \mathbf{v}^c \\
\alpha_i^j &= \frac{\exp\left(s_i^j\right)}{\sum_{k\in[1..n]} \exp\left(s_i^k\right) + \exp\left(s_i^c\right)} \\
\alpha_i^c &= \frac{\exp\left(s_i^c\right)}{\sum_{k\in[1..n]} \exp\left(s_i^k\right) + \exp\left(s_i^c\right)} \\
\mathbf{g}_i &= f_i(c, X) \\
&= \sigma\left(\sum_j \left(\alpha_i^j \mathbf{x}^j\right) + \alpha_i^c \mathbf{c}\right)
\end{aligned}
\tag{5.44}
$$

where W and b are the shared weight and bias parameters, respectively. V vectors are obtained by linear transformation of the inputs and utilized in calculating the self-attention. The unnormalized attention weight of input x_i according to input x_j is represented by s_i^j, and α_i^j is the normalized attention weight.

GSAM is used within the memory networks. To be more specific, rather than just using the context vector c, the GSAM is used for processing the inputs. The GSAMN mechanism is shown in Fig. 5.7. The value of each memory cell is updated by using the GSA as follows:

$$
\begin{aligned}
\mathbf{g}_i &= f_i\ (\mathbf{c}_k, X) \\
\mathbf{x}_i^{k+1} &= \mathbf{g}_i \odot \mathbf{x}_i^k
\end{aligned}
\tag{5.45}
$$

where $x_1^k \dots x_n^k$ is the memory value at the k-th reasoning hob. For updating the controller, both the gated self-attention and the aggregate update of the memory network are utilized. The memory state values are averaged.

$$
\begin{aligned}
\mathbf{g}_c &= f_c\ (\mathbf{c}_k, X) \\
\mathbf{c}_{k+1} &= \mathbf{g}_c \odot \mathbf{c}_k + \frac{1}{n}\sum_i \mathbf{x}_i^{k+1}
\end{aligned}
\tag{5.46}
$$

The last controller state (c_T) is considered as the representation of the given sequence. The question and the answer sentences are connected to each other and passed to the GSAMN. The first memory networks ($\mathbf{x}_1^0 \dots \mathbf{x}_n^0$) can be initialized by different embeddings including Word2Vec (Mikolov et al., 2013), GloVe (Pennington et al., 2014), ELMo (Peters et al., 2018), and BERT (Garg et al., 2020), and the control vector is initialized randomly. Then, the probability of question Q given the

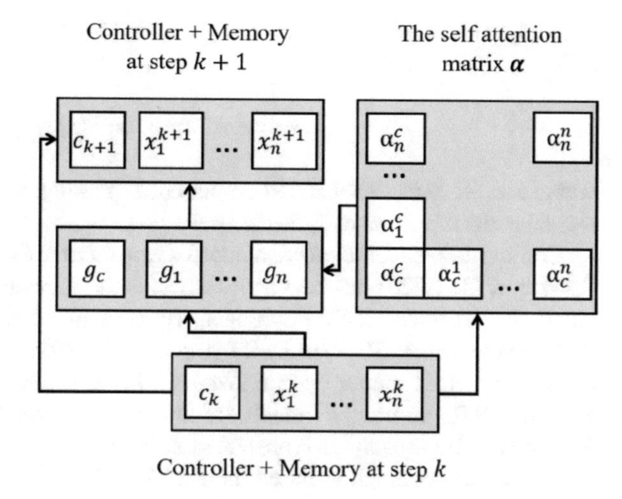

Fig. 5.7 Computation flow of GSAMN (Lai et al., 2019) ("©1963–2022 ACL, reprinted with permission")

answer sentence A is calculated by the following equation:

$$P(A|Q) = \sigma\left(\mathbf{W}_c\mathbf{c}_T + \mathbf{b}_c\right) \tag{5.47}$$

where W_c and b_c are the model parameters.

GSAMN is prepared for the answer sentence selection task by using transfer learning technique. The StackExchangeQA dataset is gathered from the Stack Exchange CQA platform and used for transfer learning. Question and answer pairs that are in English language and voted more than 2 are selected as positive pairs. Negative answers are also collected for questions. StackExchangeQA dataset includes 628,706 and 9,874,758 positive and negative question and answer pairs, respectively.

TANDA (Garg et al., 2020) uses BERT (Devlin et al., 2019) and RoBERTa (Liu et al., 2019) for solving the answer selection task. In TANDA, an additional transferring step is conducted to make the model suitable for the answer selection task. In the TANDA model, after the transferring step, TANDA is adapted to different answer selection datasets. Most of the answer selection datasets are small, and deep neural architectures need a large dataset for training. Fine-tuning BERT and RoBERTa pre-trained language models with the small size datasets will result in an unstable and noisy model. The additional transfer step in TANDA helps to build a robust to noise model and is performed over Answer Sentence Natural Question (ASNQ) dataset. ASNQ is gathered from the Natural Questions corpus (Kwiatkowski et al., 2019). Based on the architecture of TANDA, the question sentence is attached to the answer sentence with [*SEP*] token and passed to the transformer-based pre-trained language model (BERT or RoBERTa). The encoded representation of the [*CLS*] token in the beginning of the question sentence is passed to a fully connected layer followed by softMax for predicting the matching score as follows:

$$\hat{y} = \text{softMax}(W_T \times \tanh(x) + B_T) \tag{5.48}$$

where x represents the encoded output corresponding to the $[CLS]$ token, W_T is a learnable parameter, and B_T is the bias.

Contextualized Embeddings based Transformer Encoder (CETE) framework (Laskar et al., 2020) has two approaches, namely, feature-based and fine-tuning-based. In the feature-based approach, contextualized embeddings obtained from ELMo (Peters et al., 2018) or BERT (Devlin et al., 2019) are passed to a transformer encoder which is randomly initialized. In the fine-tuning-based approach, BERT and RoBERTa pre-trained models are fine-tuned on the answer selection task. The feature-based approach is a representation-based model, while the fine-tuning-based approach is an interaction-based model. The CETE framework with two different approaches is shown in Fig. 5.8.

- **Feature-based approach:** Contextualized embeddings of words are obtained from pre-trained word embeddings like BERT and ELMo. The contextualized embeddings are given to a transformer encoder. Encoder of transformer uses multi-head self-attention for generating the representation of each token based on other words within the sentence. Each self-attention mechanism generates three vectors, namely, key (K), value (V), and query (Q), for each input token by multiplying the token representation with W_Q, W_V, and W_Q weight matrices, respectively. The output vector for a given word token is generated as follows:

$$\mathrm{Z} = \text{softMax}\left(\frac{\mathrm{Q} \times \mathrm{K}^{\mathrm{T}}}{\sqrt{\mathrm{d}_k}}\right)\mathrm{V} \tag{5.49}$$

  The multi-head self-attention includes several self-attentions, and the transformer encoder which is used in CETE has eight self-attentions. The eight output Z vectors are concatenated and passed to a feed-forward layer. Mean-pooling is applied to the output of the feed-forward layer to create a fixed-size representation of the given sentence denoted by H_x and H_y. Then, the contextualized embeddings of the question and the answer sentences are compared with cosine similarity to compute the matching score.
- **Fine-tuning-based approach:** Both RoBERTa and BERT follow the same architecture for fine-tuning-based approach. Using the BERT model, two given sequences are connected with $[SEP]$ token, and $[CLS]$ token is added to the beginning of the first sequence. The aggregated representation of the given input or representation of the $[CLS]$ token is selected and passed to a classification layer with weight W. Output of classification layer is passed through a softMax layer for predicting the matching score P as follows:

$$P = \text{softMax}\left(CW^T\right) \tag{5.50}$$

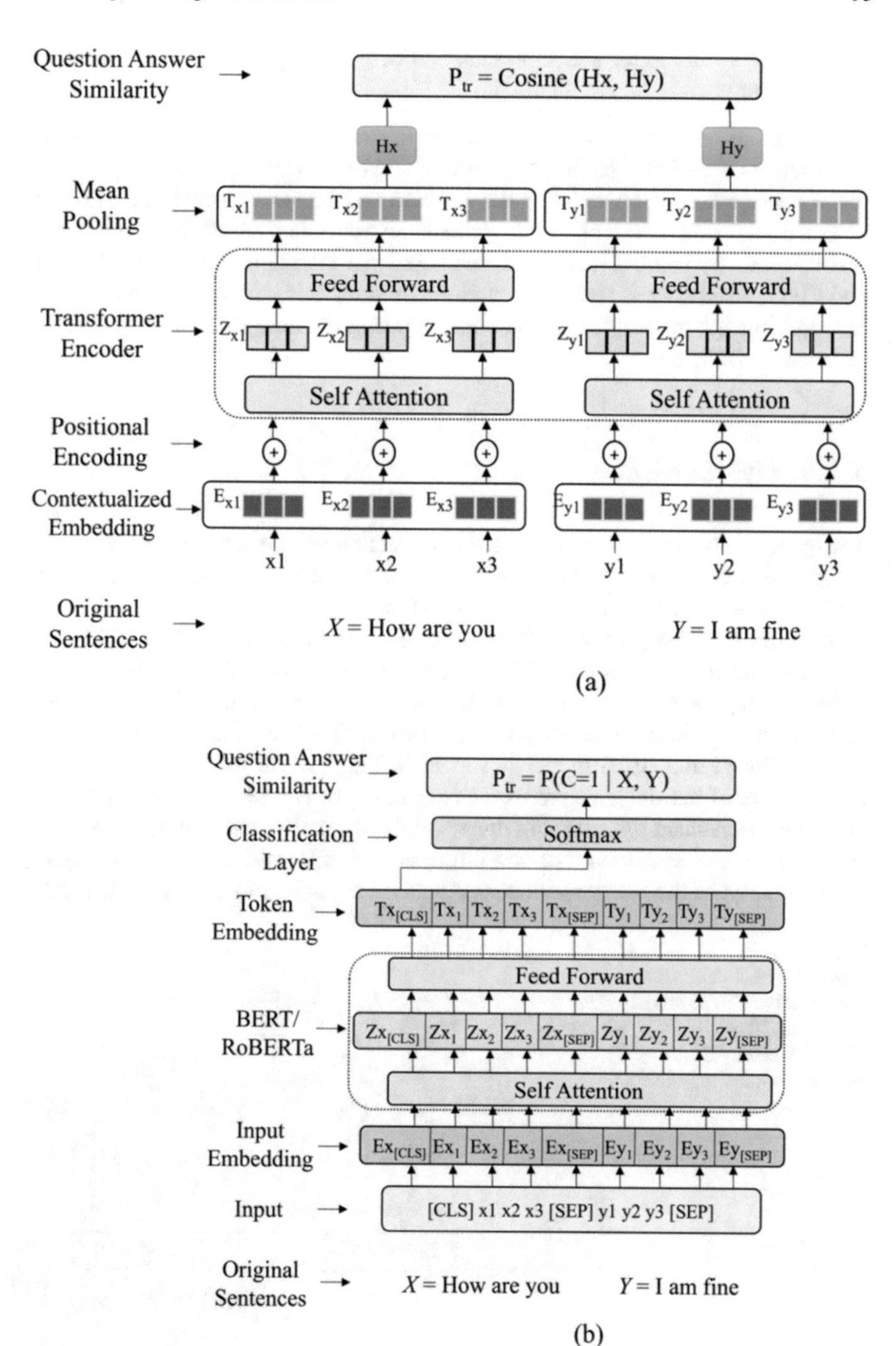

Fig. 5.8 Visualization of different approaches for CETE framework (Laskar et al., 2020): (**a**) Feature-based approach. (**b**) Fine-tuning-based approach ("©Springer Nature B.V., reprinted with permission")

where C is an aggregated representation of given input and W is a trainable parameter.

As can be seen in this section, many diverse interaction-based models are proposed. MV-LSTM, Match-SRNN, and PWIM use contextual representation of the sentence tokens for interacting the question and the answer sentences. GSAMN works pretty different from other models which form an obvious interaction matrix/tensor by using a memory-based gated self-attention mechanism. TANDA and CETE are the most recent interaction-based models, and they utilize the pre-trained language models like BERT for processing the interaction among question and answer sentences.

5.3.3 Hybrid Models

The hybrid architectures include both the interaction and the representation components. These two components can be utilized in parallel or in a sequence. We will explain some of the hybrid models in the following.

Attentive-LSTM (Tan et al., 2016) despite other variations of QA-LSTM models, including convolution-based LSTM and convolutional-pooling LSTM, does not follow the Siamese architecture and utilizes attention mechanism for generating a question-aware representation of answer sentence. The architecture of the Attentive-LSTM (Tan et al., 2016) model is shown in Fig. 5.9. An answer sentence may contain a lot of non-informative words or phrases for answering the question. The attention mechanism by assigning more weight to relative and important words or phrases reduces the influence of non-informative words or tokens in generating the representation of the answer sentence. Attention is applied to output hidden states

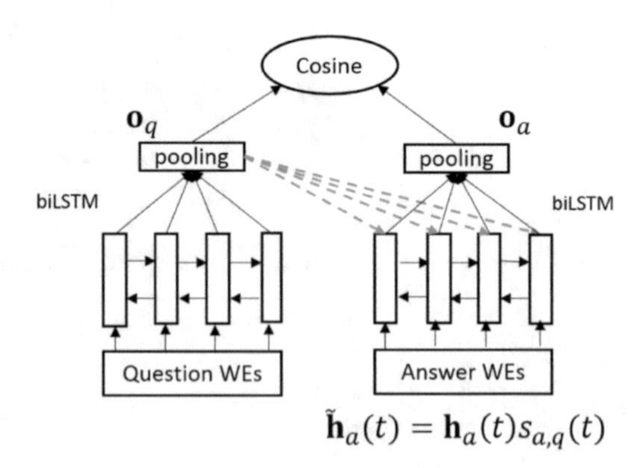

Fig. 5.9 The architecture of the Attentive-LSTM model (Tan et al., 2016) ("©1963–2022 ACL, reprinted with permission")

of answer's BiLSTM as follows:

$$
\begin{aligned}
m_{a,q}(t) &= W_{am}h_a(t) + W_{qm}o_q \\
s_{a,q}(t) &\propto \exp\left(w_{ms}^T \tanh\left(m_{a,q}(t)\right)\right) \\
\widetilde{h}_a(t) &= h_a(t)s_{a,q}(t)
\end{aligned}
\tag{5.51}
$$

where $\widetilde{h}_a(t)$ is the new question-aware representation of each BiLSTM output for tokens of answer sentence and W_{am}, W_{qm}, and w_{ms} are the attention parameters.

MCAN (Tay et al., 2018) uses attention mechanism for feature engineering. Attention mechanism is usually used as a pooling strategy which computes the attention weights for each token and uses them in generating the representation of each sequence. Then, the representations are compared and used for predicting the matching score. In MCAN, different attention mechanisms are used for extracting features from different perspectives. Features are extracted in the word level and attached to the corresponding word representation. The extracted features contain useful information about the relation of question and answer sentences. By using three distinct co-attentions and one intra-attention mechanism, four attentional features are obtained for each word. New augmented representation of words is passed to an LSTM network for modeling the contextual relation of the question and the answer sentences efficiently.Each component of this model is described below.

- **Input encoder:** Highway encoder layer is used for encoding the input word embeddings $w \in R^d$ where d is the dimensionality of word embedding. Highway encoder, similar to RNNs, controls the flow of information to the next layers, and they can filter non-informative words. Formulation of a highway encoder with one layer is as follows:

$$
y = H\left(x, W_H\right) \cdot T\left(x, W_T\right) + \left(1 - T\left(x, W_T\right)\right) \cdot x \tag{5.52}
$$

  where H and T are one-layer affine transformers with ReLU and sigmoid activation functions, respectively. $W_H and W_T \in R^{r \times d}$ are the model parameters.
- **Co-attention:** Four different types of co-attention mechanism, namely, (1) max-pooling, (2) mean-pooling, (3) alignment-pooling, and (4) intra-attention (or self-attention), are used. In co-attention, interaction between each pair from two sequences is modeled by creating a similarity matrix. Here, the similarity matrix is built as follows:

$$
s_{ij} = F\left(q_i\right)^T F\left(d_j\right) \tag{5.53}
$$

where F is a MLP. The similarity matrix can also be calculated by the following equations:

$$s_{ij} = q_i^T M d_j \tag{5.54}$$

$$s_{ij} = F\left[q_i; d_j\right]\Big] \tag{5.55}$$

Having the similarity matrix S, each of the mentioned variants of co-attention mechanism is calculated as follows:

1. **Extractive pooling:** Extractive pooling has two variants including max-pooling and mean-pooling. In the max-pooling approach, each word is modeled according to its highest importance or its highest influence on each word in the other sentence, while in the mean-pooling approach, the overall influence of a word on words of the other sequence is considered. Formulations of the both approaches are defined as follows:

$$\begin{aligned} q' &= Soft\left(\max_{col}(s)\right)^{\top} q \quad and \quad d' = Soft\left(\max_{row}(s)\right)^{\top} d \\ q' &= Soft(mean(s))^{\top} q \quad and \quad d' = Soft\left(mean_{row}(s)\right)^{\top} d \end{aligned} \tag{5.56}$$

   where q' and d' are co-attentive representations of the question and document.
2. **Alignment pooling:** Co-attentive representations of each subphrase in question q_i' and document d_i' are learned as follows:

$$d_i' := \sum_{j=1}^{\ell_q} \frac{\exp\left(s_{ij}\right)}{\sum_{k=1}^{\ell_q} \exp\left(s_{ik}\right)} q_j \quad and \quad q_j' := \sum_{i=1}^{\ell_d} \frac{\exp\left(s_{ij}\right)}{\sum_{k=1}^{\ell_d} \exp\left(s_{kj}\right)} d_i \tag{5.57}$$

   where d_i' indicates a subphrase in q which is softly aligned to d_i or d_i' is the weighted summation of q terms where the relevant words to q_i have more weight.
3. **Intra-attention:** Long-term dependencies within a sentence are influential in creating intra-attention of or self-attention. Intra-attention is employed on one sequence, and in this model, it is employed on question and document sentences separately as follows:

$$x_i' := \sum_{j=1}^{\ell} \frac{\exp\left(s_{ij}\right)}{\sum_{k=1}^{\ell} \exp\left(s_{ik}\right)} x_j \tag{5.58}$$

   where x can be each of the q or d and x_i' indicates the self-attentional representation of x_j.

- **Multi-cast attention:** The attentional representation, generated for each token, is compared with its original representation from three different perspectives to extract features. Each feature represents the impact of attention on pure representation. Three similarity measurement functions are used for comparing a token x with its attentional representation $\overline{x}$ as follows:

$$
\begin{aligned}
f_c &= F_c([\overline{x}; x]) \\
f_m &= F_c(\overline{x} \odot x) \\
f_s &= F_c(\overline{x} - x)
\end{aligned}
\tag{5.59}
$$

  where $\odot$ represents the Hadamard product and ; concatenates two vectors. F_c indicates a function which compresses the given vector to a scalar value. Three comparison functions including sum (SM), neural network (NN), and factorization machines (FM) can be used for converting a vector to a scalar. For each term, four attentional representations are produced using the four co-attentions explained above, and each of them is converted to three scalar values using three various compression functions. Therefore, the feature vector $z \in R^{12}$ is obtained for each term and concatenated with word representation w_i to form the augmented vector representation $\bar{w}_i = [w_i; z_i]$.
- **LSTM encoder:** An LSTM is used for modeling the long-term dependencies among the question and document terms separately. The augmented representation given to LSTM helps it by providing additional information about long-term and short-term dependencies within the sequence and between two sequences. The output hidden states of LSTM (h_i) are pooled by using *MeanMax* pooling which concatenates the result of *Mean* and *Max* pooling approaches. Each of the query and the document is modeled separately in this layer.

$$
\begin{aligned}
\mathrm{H_i} &= \mathrm{LSTM(u, i)}, \quad \forall \mathrm{i} \in [1, 2, \ldots, 1] \\
\mathrm{H} &= \mathrm{MeanMax}\,[\mathrm{h}_1 \ldots \mathrm{h}_l]
\end{aligned}
\tag{5.60}
$$

- **Prediction layer and optimization:** Representations of the question and document sentences generated in the previous layer (x_q and x_d) are passed through two-layer highway encoder as follows:

$$
y_{out} = H_2\left(H_1\left(\left[x_q; x_d; x_q \odot x_d; x_q - x_d\right]\right)\right) \tag{5.61}
$$

  where H_1 and H_2 are highway encoders followed by ReLU activation functions. Output of highway encoder layer is given to feed-forward layer followed by softMax layer for predicting the matching score y_{pred} as follows:

$$
y_{pred} = softMax\,(W_F \cdot y_{out} + b_F) \tag{5.62}
$$

  where W_F and b_F are the model parameter and the bias, respectively.

Bian et al. (2017) improved the compare-aggregate model by proposing a new type of attention mechanism. The answer sentence can be very long and can contain a lot of irrelevant terms for answering the question. In the usual attention mechanism, the attention weights for each token are calculated, and then the sentence representation is created by averaging the weighted representation of the tokens according to the attention weights. Even though the usual attention mechanism detects the irrelevant tokens and gives them a low weight, a large number of small weights accumulate and bring noise to the representation. For solving the mentioned problem, two dynamic-clip attentions, namely, k-max and k-threshold, are proposed. The compare-aggregate model is trained with a list-wise approach which best suits the nature of the ranking task. Components of this model are explained in the following.

- **Word representation layer:** Question and answer sentences are shown as a sequence of their word embeddings $\overrightarrow{\mathrm{q}} = (\mathrm{q}_1, \ldots, \mathrm{q}_{\ell_q})$ and $\overrightarrow{\mathrm{a}} = (\mathrm{a}_1, \ldots, \mathrm{a}_{\ell_a})$, respectively.
- **Attention layer:** Attention mechanism is used for creating the soft-aligned representation of the given sentences. The attention matrix is built by computing the dot product similarity of each token from two sequences as follows:

$$e_{ij} = q_i \cdot a_j \tag{5.63}$$

where q_i and a_j are i-th question and j-th answer terms, respectively. The attention weights w_{ij}^a and w_{ij}^q are calculated as follows:

$$\begin{aligned} w_{ij}^a &= \frac{\exp\left(e_{ij}\right)}{\sum_{k=1}^{\ell_q} \exp\left(e_{kj}\right)}, \\ w_{ij}^q &= \frac{\exp\left(e_{ij}\right)}{\sum_{k=1}^{\ell_a} \exp\left(e_{ik}\right)} \end{aligned} \tag{5.64}$$

The attention weights are filtered with two different filtering approaches to eliminate the effect of noisy and irrelevant tokens in computations. These two approaches are explained below:

 - *K*-**max attention:** Attention weights w_{ij} are sorted with descending order, and the first k weights are preserved in set S, while the other ones are changed to zero. This approach by limiting the number of influential tokens eliminated the noise.

$$\begin{cases} w_{ij} = \frac{w_{ij}}{\sum_i^S w_{ij}} & i \in S \\ w_{ij} = 0 & j \notin S \end{cases} \tag{5.65}$$

 - *K*-**threshold attention:** Attention weights larger than the specified threshold value k are preserved in set S, and the weights less than this value are changed to zero. A threshold for relevant interactions tokens is determined, and all of the relevant tokens are considered in computations.

$$\begin{cases} w_{ij} = w_{ij} & w_{ij} \geq k \\ w_{ij} = 0 & w_{ij} < k \end{cases} \tag{5.66}$$

After determining the attention weights, they are normalized and used for creating the vector representation of the question sentence h_j^a which is aligned to the answer term j and vector representation of the answer sentence h_j^q which is aligned to the question term j as follows:

$$w_{ij} = \frac{w_{ij}}{\sum_{k=1}^{\ell_q} w_{ij}} \tag{5.67}$$

$$h_j^a = \sum_{i=1}^{\ell_q} w_{ij}^a q_i, \quad h_i^q = \sum_{j=1}^{\ell_a} w_{ij}^q a_j \tag{5.68}$$

- **Comparison:** Element-wise multiplication is used for comparison. Each question term q_i is compared with representation of the answer sentence aligned with that question term h_i^q as follows:

$$t_j^a = f\left(a_j, h_j^a\right) = a_j \otimes h_j^a \tag{5.69}$$

where $\otimes$ indicates the element-wise multiplication.
- **Aggregation:** Results of the comparison layer, t_j^a and t_j^q, are aggregated using a CNN with one layer. The aggregated vectors are concatenated and used for predicting the matching score as follows:

$$r_a = CNN\left(\left[t_1^a, \ldots, t_{l_a}^a\right]\right), \quad r_q = CNN\left(\left[t_1^q, \ldots, t_{l_q}^q\right]\right) \tag{5.70}$$

$$Score = \left[r_a, r_q\right]^T W \tag{5.71}$$

where W is the model parameter.
- **List-wise learning to rank:** Question Q with a set of candidate answers $A = \{A_1, A_2 \ldots A_N\}$ with true labels $Y = \{y_1, y_2 \ldots y_N\}$ is fed to the model, and a set of scores S are obtained and normalized by softMax function. The true labels are also normalized. Finally, the KL divergence loss is utilized for training the model as follows:

$$Score_j = model\left[\mathbf{Q}, \mathbf{A}_j\right] \tag{5.72}$$

$$\mathrm{S} = softMax\left(\left[Score_1, \ldots, Score_N\right]\right) \tag{5.73}$$

$$\mathrm{Y} = \frac{\mathrm{Y}}{\sum_{i=1}^{N} y_i} \tag{5.74}$$

$$loss = \frac{1}{n}\sum_{1}^{n} KL(\,\mathrm{S}\|\mathrm{Y}) \tag{5.75}$$

Comp-Clip model (Yoon et al., 2019) has a compare-aggregate architecture. ELMo pre-trained word embeddings are used instead of the word representation layer. The pre-trained language models represent the contextual information of the words more efficiently. Transfer learning technique is used for mitigating the small size of the datasets and preparing the model for answer sentence selection. Question-answering Natural Language Inference (QNLI) dataset (Wang et al., 2018) is used for transfer learning. A novel latent clustering model is used for extracting additional features. The latent cluster information of each sentence is computed by comparing it with latent memory. A pair of a question and a true answer has similar latent information. This auxiliary information obtained from the latent clustering model helps to better model the question and the answer pair similarity. Both point-wise and pair-wise learning to rank approaches are used for training in the objective function. The point-wise learning to rank approach outperforms the pair-wise approach in the conducted experiments. The matching score $f(y|Q, A)$ or probability of label y, given the question Q and the answer A sentences, is computed in the output of Comp-Clip. The architecture of Comp-Clip includes six layers that are explained in the following.

- **Pre-trained language model:** In the case of using the ELMo pre-trained language model instead of the word embedding layer, the input sentences Q and A are replaced by $L^Q = ELMo(Q)$ and $L^{\mathrm{A}} = ELMo(A)$ in the following equations.
- **Context representation:** Context information of the question ($Q \in R^{d\times Q}$) and the answer ($A \in R^{d\times A}$) sequences with length Q and A, respectively, are captured by:

$$\begin{aligned}\overline{\mathbf{Q}} &= \sigma\left(\mathbf{W}^i\mathbf{Q}\right) \odot \tanh\left(\mathbf{W}^u\mathbf{Q}\right)\\ \overline{\mathbf{A}} &= \sigma\left(\mathbf{W}^i\mathbf{A}\right) \odot \tanh\left(\mathbf{W}^u\mathbf{A}\right)\end{aligned} \tag{5.76}$$

where $\odot$ is the element-wise multiplication, σ is the sigma function, $W \in R^{l\times d}$ is the model parameter, and d is the dimensionality of word embedding.

- **Attention:** Dynamic-clip attention is applied on context representation of each sentence to create soft alignment of question $H^Q R^{l\times Q}$ and answer $H^A R^{l\times A}$ as follows:

$$\mathbf{H}^Q = \overline{\mathbf{Q}} \cdot softMax\left(\left(\mathbf{W}^q \overline{\mathbf{Q}}\right)^\top \overline{\mathbf{A}}\right)$$
$$\mathbf{H}^A = \overline{\mathbf{A}} \cdot softMax\left(\left(\mathbf{W}^a \overline{\mathbf{A}}\right)^\top \overline{\mathbf{Q}}\right) \tag{5.77}$$

- **Comparison:** Attentional representation of each sequence is compared with representation of the other sequence as follows:

$$\mathbf{C}^Q = \overline{\mathbf{A}} \odot \mathbf{H}^Q, \left(\mathbf{C}^Q \in R^{l\times A}\right)$$
$$\mathbf{C}^A = \overline{\mathbf{Q}} \odot \mathbf{H}^A, \left(\mathbf{C}^A \in R^{l\times Q}\right) \tag{5.78}$$

- **Aggregation:** Output of the comparison layer is aggregated by using CNN (with n types of filter), and the aggregated vectors are used for calculating the matching score of question and answer sentences as follows:

$$\mathbf{R}^Q = \mathrm{CNN}\left(\mathbf{C}^Q\right), \mathbf{R}^A = \mathrm{CNN}\left(\mathbf{C}^A\right)$$
$$score = \sigma\left(\left[\mathbf{R}^Q; \mathbf{R}^A\right]^\top \mathbf{W}\right) \tag{5.79}$$

where $\mathbf{W} \in \mathbb{R}^{2nl\times 1}$ is the model parameter and ; concatenates two vectors $\mathbf{R}^Q \in \mathbb{R}^{nl}$ and $\mathbf{R}^A \in \mathbb{R}^{nl}$.

- **Latent clustering:** Latent memory networks are created, and similarity of the given sentence with latent memories is calculated to obtain the latent cluster information. The LS information of each sentence is the weighted sum of the latent memories based on their similarity as follows:

$$\mathbf{p}_{1:n} = \mathbf{s}^\top \mathbf{W} \mathbf{M}_{1:n}$$
$$\overline{\mathbf{p}}_{1:k} = k - \max - pool\left(\mathbf{p}_{1:n}\right)$$
$$\alpha_{1:k} = softMax\left(\overline{\mathbf{p}}_{1:k}\right)$$
$$\mathbf{M}_{\mathrm{LC}} = \Sigma_k \bar{\alpha}_k \mathbf{M}_k \tag{5.80}$$

where $M_{1:n} \in R^{\prime d\times n}$ is the latent memory, $s \in R^d$ is the sentence representation, and $\mathbf{W} \in \mathbb{R}^{d\times d'}$ is the model parameter.

The latent information of each sentence is calculated and added to the result of comparison layer as an auxiliary feature as follows:

$$\mathbf{M}_{\mathrm{LC}}^{Q} = f\left(\left(\Sigma_i \bar{q}_i\right)/n\right), \bar{q}_i \subset \overline{\mathbf{Q}}_{1:n}$$
$$\mathbf{M}_{\mathrm{LC}}^{A} = f\left(\left(\Sigma_i \bar{a}_i\right)/m\right), \bar{a}_i \subset \overline{\mathbf{A}}_{1:m} \quad (5.81)$$
$$\mathbf{C}_{new}^{Q} = \left[\mathbf{C}^{Q}; \mathbf{M}_{\mathrm{LC}}^{Q}\right], \mathbf{C}_{new}^{A} = \left[\mathbf{C}^{A}; \mathbf{M}_{\mathrm{LC}}^{A}\right]$$

where f indicates the latent clustering model. The sentence representation is considered as average of sequence tokens, and the new obtained representations for the question C_{new}^{Q} and the answer C_{new}^{A} sentences replace the pure representation C^{Q} and C^{A} in the above formulas.

RE2 (Yang et al., 2019) is developed by the main intuition of developing a fast and simple text matching model by using lightweight and simple layers for aligning the sequences. The architecture of the RE2 model is shown in Fig. 5.10. The embedding layer embeds the tokens within the sequence. The given sequence is processed by several similar blocks which contain encoding, fusion, and alignment layers. The similar blocks are connected to each other by residual connections. The term RE2 stands for residual vectors (rectangles with diagonal stripes generated in the previous blocks), embedding vectors (white rectangles generated in the embedding layer), and encoded vectors (black rectangles generated by the encoder layer). According to the architecture of RE2, which is shown in Fig. 5.10, RE2 includes two separate architectures for processing each sequence that are connected

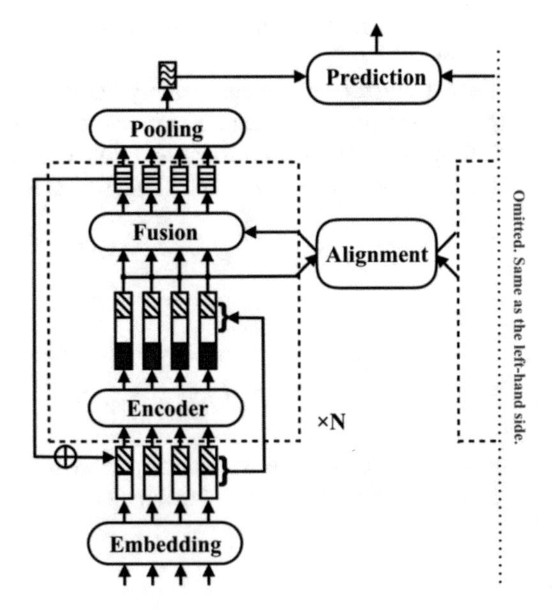

Fig. 5.10 The architecture of the RE2 model (Yang et al., 2019) ("©1963–2022 ACL, reprinted with permission")

with the alignment layer. A sequence representation is created in the pooling layer given the sequential output of the fusion layer. The two representations generated in each part of the model are compared in the prediction layer. The main components in the architecture of RE2 are explained in the following.

- **Augmented residual connections:** The residual connections are used for extracting rich features from the text sequence. Embedding of the words is passed to the first block, while the input of the next blocks ($n > 1$) is augmented by using the summation of the output of the two previous blocks. Therefore, input of the i-th block ($x^{(n)}$) is generated according to:

$$x_i^{(n)} = \left[x_i^{(1)}; o_i^{(n-1)} + o_i^{(n-2)}\right] \tag{5.82}$$

where $o^{(n-1)}$ is the output of the n-th block, $o^{(n)} = \left(o_1^{(n)}, o_2^{(n)}, \ldots, o_l^{(n)}\right)$, $x^{(n)} = \left(x_1^{(n)}, x_2^{(n)}, \ldots, x_l^{(n)}\right)$, and ; indicates their concatenation.
- **Alignment layer:** For calculating the aligned representation of the input sequences $a = \left(a_1, a_2, \ldots, a_{l_a}\right)$ and $b = \left(b_1, b_2, \ldots, b_{l_b}\right)$, where l_x is the length of the sequence x, the similarity between two terms of the first and the second sequences, a_i and b_j, respectively, is calculated as follows:

$$e_{ij} = F\left(a_i\right)^T F\left(b_j\right) \tag{5.83}$$

where F is a feed-forward neural network with one layer.

Aligned representation of the given sequences, namely, a_i' and b_i', is calculated by summing the weighted representation of the other sentence according to attention weights as follows:

$$\begin{aligned} a_i' &= \sum_{j=1}^{l_b} \frac{\exp\left(e_{ij}\right)}{\sum_{k=1}^{l_b} \exp\left(e_{ik}\right)} b_j \\ b_j' &= \sum_{i=1}^{l_a} \frac{\exp\left(e_{ij}\right)}{\sum_{k=1}^{l_a} \exp\left(e_{kj}\right)} a_i \end{aligned} \tag{5.84}$$

- **Fusion layer:** The local and the aligned representations are compared from different perspectives, and the obtained results are combined. In all of these perspectives, a feed-forward neural network is used for comparison. The results of comparison $\bar{a}_i^1$, $\bar{a}_i^2$, and $\bar{a}_i^3 = G_3$ for sequence a are calculated as follows:

$$\begin{aligned} \bar{a}_i^1 &= G_1\left(\left[a_i; a_i'\right]\right) \\ \bar{a}_i^2 &= G_2\left(\left[a_i; a_i - a_i'\right]\right) \\ \bar{a}_i^3 &= G_3\left(\left[a_i; a_i \circ a_i'\right]\right) \end{aligned} \tag{5.85}$$

where G_1, G_2, and G_3 are the feed-forward neural networks with one layer and $\circ$ is the concatenation operator. Outputs of the comparison functions are combined by the following equation:

$$\bar{a}_i = G\left(\left[\bar{a}_i^1; \bar{a}_i^2; \bar{a}_i^3\right]\right) \quad (5.86)$$

where G is also a one-layer feed-forward neural network. The output of the fusion layer for sequence b, named $\bar{b}$, is calculated accordingly.

- **Prediction layer:** The vector representation of the two given sequences, namely, v_1 and v_2, is obtained from the pooling layer and used for predicting the output as follows:

$$\hat{\mathbf{y}} = H\left([v_1; v_2; v_1 - v_2; v_1 \circ v_2]\right) \quad (5.87)$$

where $\hat{\mathbf{y}} \in \mathcal{R}^C$ is the predicted score for each class, C is the count of classes in a multi-class task, and H is the multi-layer feed-forward neural network. The predicted class is $\hat{y} = argmax_i \hat{\mathbf{y}}_i$.

Bilateral Multi-Perspective Matching Model (BiMPM) (Wang et al., 2017) has compare-aggregate architecture and tries to improve this architecture by matching sequences in two different directions and exploring phrase- and sentence-level matching rather than just word-level matching. Each of the given sentences P and Q is encoded using a BiLSTM. The encoded representations are matched in two directions $P \rightarrow Q$ and $P \leftarrow Q$. For matching $P \rightarrow Q$, each hidden state of Q is matched from different perspectives with all hidden states of P. Results of the matching layer are aggregated to generate the matching vectors. Matching vectors are compared to predict the probability of label y given the two input sentences P and Q ($Pr(y|P, Q)$). BiMPM includes five major layers that are described below.

- **Word representation layer:** Representation of each token is built by both the word and the character-composed embeddings. Word embeddings are obtained from pre-trained GloVe (Pennington et al., 2014) or Word2Vec (Mikolov et al., 2013) embeddings. The character-composed embedding is obtained by using an LSTM (Hochreiter & Schmidhuber, 1997). Each of the sentences P and Q is represented by a set of d-dimensional vectors $P : \left[\boldsymbol{p}_1, \ldots, \boldsymbol{p}_M\right]$ and $Q : \left[\boldsymbol{q}_1, \ldots, \boldsymbol{q}_N\right]$, respectively.
- **Context representation layer:** By using BiLSTM, context information of each token of P and Q is generated as follows:

$$\begin{aligned} \overrightarrow{\boldsymbol{h}}_i^p &= \overrightarrow{LSTM}\left(\overrightarrow{\boldsymbol{h}}_{i-1}^p, \boldsymbol{p}_i\right) \quad i = 1, \ldots, M \\ \overleftarrow{\boldsymbol{h}}_i^p &= \overleftarrow{\text{LSTM}}\left(\overleftarrow{\boldsymbol{h}}_{i+1}^p, \boldsymbol{p}_i\right) \quad i = M, \ldots, 1 \end{aligned} \quad (5.88)$$

$$\overrightarrow{\boldsymbol{h}}_j^q = \overrightarrow{LSTM}\left(\overrightarrow{\boldsymbol{h}}_{j-1}^q, \boldsymbol{q}_j\right) \quad j = 1, \ldots, N$$
$$\overleftarrow{\boldsymbol{h}}_j^q = \overleftarrow{LSTM}\left(\overleftarrow{\boldsymbol{h}}_{j+1}^q, \boldsymbol{q}_j\right) \quad j = N, \ldots, 1 \tag{5.89}$$

- **Matching layer:** A multi-perspective matching model is proposed to match two given sequences. Two given sequences are matched in two directions $P : \left[\boldsymbol{p}_1, \ldots, \boldsymbol{p}_M\right]$ and $Q : \left[\boldsymbol{q}_1, \ldots, \boldsymbol{q}_N\right]$. In each direction, every time step from one sequence is compared with all time steps of the other sequence. The multi-perspective cosine matching function f_m is defined as below:

$$\boldsymbol{m} = f_m\left(\boldsymbol{v}_1, \boldsymbol{v}_2; \boldsymbol{W}\right) \tag{5.90}$$

where v_1 and v_2 are the d-dimensional input vectors and $W \in R^{l \times d}$ is the model parameter where l is the number of perspectives. Each element of returned vector $m \in R^l$ is the result of comparison from k-th perspective as follows:

$$m_k = cosine\left(W_k \circ \boldsymbol{v}_1, W_k \circ \boldsymbol{v}_2\right) \tag{5.91}$$

where W_k is the k-th row of parameter W or k-th perspective.

Four different matching strategies are proposed based on the f_m function. We will explain the matching strategies (for one direction $P \rightarrow Q$) in the following.

1. Full-Matching: Each time step, which is the contextualized representation, from one sequence is compared with the last time step of the other sequence in the forward and backward directions as follows:

$$\overrightarrow{\boldsymbol{m}}_i^{full} = f_m\left(\overrightarrow{\boldsymbol{h}}_i^p, \overrightarrow{\boldsymbol{h}}_N^q; \mathbf{W}^1\right)$$
$$\overleftarrow{\boldsymbol{m}}_i^{full} = f_m\left(\overleftarrow{\boldsymbol{h}}_i^p, \overleftarrow{\boldsymbol{h}}_1^q; \mathbf{W}^2\right) \tag{5.92}$$

where $\overrightarrow{\boldsymbol{h}}_i^p$ and $\overleftarrow{\boldsymbol{h}}_i^p$ are the i-th forward and backward time steps of sequence P, respectively.

2. Maxpooling-Matching: Each time step representation is compared with all time steps of the other sentence, and the maximum value of each dimension is preserved. Maxpooling-Matching is calculated for forward and backward time steps as follows:

$$\overrightarrow{\boldsymbol{m}}_i^{max} = \max_{j \in (1 \ldots N)} f_m\left(\overrightarrow{\boldsymbol{h}}_i^p, \overrightarrow{\boldsymbol{h}}_j^q; \mathbf{W}^3\right)$$
$$\overleftarrow{\boldsymbol{m}}_i^{max} = \max_{j \in (1 \ldots N)} f_m\left(\overleftarrow{\boldsymbol{h}}_i^p, \overleftarrow{\boldsymbol{h}}_j^q; \mathbf{W}^4\right) \tag{5.93}$$

3. Attentive-Matching: Similarity of each time step is computed with all time steps of the other sentence by cosine similarity for each of the forward and backward directions as follows:

$$\begin{aligned}\overrightarrow{\alpha}_{i,j} &= cosine\left(\overrightarrow{\boldsymbol{h}}_i^p, \overrightarrow{\boldsymbol{h}}_j^q\right) \quad j = 1, \ldots, N \\ \overleftarrow{\alpha}_{i,j} &= cosine\left(\overleftarrow{\boldsymbol{h}}_i^p, \overleftarrow{\boldsymbol{h}}_j^q\right) \quad j = 1, \ldots, N\end{aligned} \tag{5.94}$$

A weighted representation of sentence Q is generated by using the above attention weights as follows:

$$\begin{aligned}\overrightarrow{\boldsymbol{h}}_i^{mean} &= \frac{\sum_{j=1}^{N} \boldsymbol{\alpha}_{i,j} \cdot \overrightarrow{\boldsymbol{h}}_j^q}{\sum_{j=1}^{N} \boldsymbol{\alpha}_{i,j}} \\ \overleftarrow{\boldsymbol{h}}_i^{mean} &= \frac{\sum_{j=1}^{N} \overleftarrow{\alpha}_{i,j} \cdot \overleftarrow{\boldsymbol{h}}_j^q}{\sum_{j=1}^{N} \overleftarrow{\alpha}_{i,j}}\end{aligned} \tag{5.95}$$

The attentional representation of the sentence Q in forward $\overrightarrow{\boldsymbol{h}}_i^{mean}$ and backward $\overleftarrow{\boldsymbol{h}}_i^{mean}$ directions is used for comparing with $\overrightarrow{\boldsymbol{h}}_i^p$ and $\overleftarrow{\boldsymbol{h}}_i^p$

$$\begin{aligned}\overrightarrow{\boldsymbol{m}}_i^{att} &= f_m\left(\overrightarrow{\boldsymbol{h}}_i^p, \overrightarrow{\boldsymbol{h}}_i^{mean}; \mathbf{W}^5\right) \\ \overleftarrow{\boldsymbol{m}}_i^{att} &= f_m\left(\overleftarrow{\boldsymbol{h}}_i^p, \overleftarrow{\boldsymbol{h}}_i^{mean}; \mathbf{W}^6\right)\end{aligned} \tag{5.96}$$

4. Max-Attentive-Matching: This strategy is similar to Attentive-Matching, but instead of computing the weighted sum of hidden vectors, the hidden vector with the highest cosine similarity is chosen as attentive vector.

All of the above matching strategies are applied to each hidden vector in both directions, and eight outputs are concatenated to form the result of the matching layer.

- **Aggregation layer:** Output of each matching layer is given to a BiLSTM for aggregating them to a fixed-size vector representation. The last hidden state of each LSTM is concatenated to form the aggregation result.
- **Prediction layer:** Output of the aggregation layer for each sequence is concatenated and passed to a two-layer feed-forward neural network for predicting the probability $Pr(y|P, Q)$.

Wang and Jiang (2017) proposed a model with compare-aggregate architecture. Various comparison functions are used in the comparison layer. The model includes four layers that are described in the following.

- **Preprocessing:** Contextualized representation of each token is generated by using a modified version of LSTM/GRU. In the modified version, just the input gate is preserved to remember informative words. Given the question sentence Q and answer sentence A, their contextualized representations $\bar{Q}$ and $\bar{A}$ are computed as follows:

$$\overline{Q} = \sigma\left(W^{\mathrm{i}}Q + b^{i} \otimes e_{Q}\right) \odot \tanh\left(W^{\mathrm{u}}Q + b^{\mathrm{u}} \otimes e_{Q}\right)$$
$$\overline{A} = \sigma\left(W^{\mathrm{i}}A + b^{\mathrm{i}} \otimes e_{A}\right) \odot \tanh\left(W^{\mathrm{u}}A + b^{\mathrm{u}} \otimes e_{A}\right) \tag{5.97}$$

where $W^{i} \in R^{l\times d}$ and $W^{u} \in R^{l\times d}$ are the model parameters and $b^{i} \in R^{l}$ and $b^{u} \in R^{l}$ are the biases. $\odot$ is the element-wise multiplication, and $(\cdot \otimes \mathbf{e}_{X})$, by repeating the vector or scalar on the left for x time, creates a matrix or vector, respectively.

- **Attention:** Attention weight for each column of answer's contextualized representation $\overline{A}$ is generated over each column vector of $\overline{Q}$. Using the attention weights, vector h_j is produced for each column vector of $\overline{A}$ (or contextualized representation of j-th term $\overline{a}_j$) as follows:

$$G = softMax\left(\left(W^{g}\overline{Q} + b^{g} \otimes e_{Q}\right)^{\mathrm{T}} A\right)$$
$$H = \overline{Q}G \tag{5.98}$$

where $\mathbf{W}^{g} \in \mathbb{R}^{l\times l}$ and $\mathbf{b}^{g} \in \mathbb{R}^{l}$ are the model parameters, $\mathbf{G} \in \mathbb{R}^{Q\times A}$ is the matrix of attention weights, and each column of $\mathbf{H} \in \mathbb{R}^{l\times A}$ denoted as h_j is the weighted sum of the question words or column vectors of $\bar{Q}$.

- **Comparison:** Contextualized representation of each answer term $\overline{a}_j$ is compared with the result of attention layer h_j. h_j is the representation of the question sentence that best matches the answer term $\overline{a}_j$. There are multiple comparison functions denoted as f for comparison of these two vectors that are explained in the following.

1. Neural Net (NN):

$$t_j = f\left(\overline{a}_j, h_j\right) = ReLU\left(W\begin{bmatrix} a_j \\ h_j \end{bmatrix} + b\right) \tag{5.99}$$

where $\mathbf{W} \in \mathbb{R}^{l\times 2l}$ and $\mathbf{b} \in \mathbb{R}^{l}$ are the model parameters.

2. Neural Tensor Net (NTN):

$$t_j = f\left(\overline{a}_j, h_j\right) = ReLU\left(a_j^{\mathrm{T}} T^{[1\ldots l]} h_j + b\right) \tag{5.100}$$

where $\mathbf{T}^{[1\ldots l]} \in \mathbb{R}^{l\times l\times l}$ and $\mathbf{b} \in \mathbb{R}^{l}$ are the model parameters.

3. Euclidean + Cosine (EUCCOS):

$$t_j = f\left(\overline{a}_j, h_j\right) = \begin{bmatrix} \left\|a_j - h_j\right\|_2 \\ \cos\left(\overline{a}_j, h_j\right) \end{bmatrix} \tag{5.101}$$

As the output of EUCCOS comparison function is just a two-dimensional vector, it may omit some useful information about two given sequences. On the contrary, the neural network-based functions are too general and cannot represent the similarity of two vectors efficiently. The following comparison functions are proposed to alleviate the mentioned problems.

4. Subtraction (SUB): SUB is similar to Euclidean distance, and by summing elements of its output, the Euclidean distance between two given vectors is generated. Information about various dimensions of two given vectors are preserved in the SUB function.

$$t_j = f\left(\overline{a}_j, h_j\right) = \left(\overline{a}_j - h_j\right) \odot \left(\overline{a}_j - h_j\right) \tag{5.102}$$

5. Multiplication (MULT): MULT is analogous to cosine similarity, while it maintains some information about the given vectors.

$$t_j = f\left(\overline{a}_j, h_j\right) = \overline{a}_j \odot h_j \tag{5.103}$$

6. SUBMULT+NN: It is a combination of MULT and SUB functions followed by a feed-forward neural network.

$$t_j = f\left(\overline{a}_j, h_j\right) = ReLU\left(W\begin{bmatrix} \left(\overline{a}_j - h_j\right) \odot \left(\overline{a}_j - h_j\right) \\ \overline{a}_j \odot h_j \end{bmatrix} + b\right) \tag{5.104}$$

- **Aggregation:** A series of t_i vectors are produced in the comparison layer by comparing contextualized representation of each answer term $\overline{\mathbf{a}}_j$ with h_j. These vectors are aggregated to a fixed-size vector $\mathbf{r} \in \mathbb{R}^{nl}$ by applying a CNN over them as follows:

$$r = \text{CNN}\left(\left[t_1, \ldots, t_A\right]\right) \tag{5.105}$$

where n is the count of convolution windows. The output vector is used in the prediction layer for predicting the matching score.

Yin et al. (2016) proposed three different models based on BCNN, which replace the average pooling with attentional pooling. The architecture of the ABCNN-1 model is shown in Fig. 5.11. Matrix A is built by comparing two given sentences and is used for creating weighted representation of one sentence according to its relevance to the other sentence. Each row of matrix A represents the attention weight of corresponding unit in the first sentence (s_0) according to the second sentence (s_1),

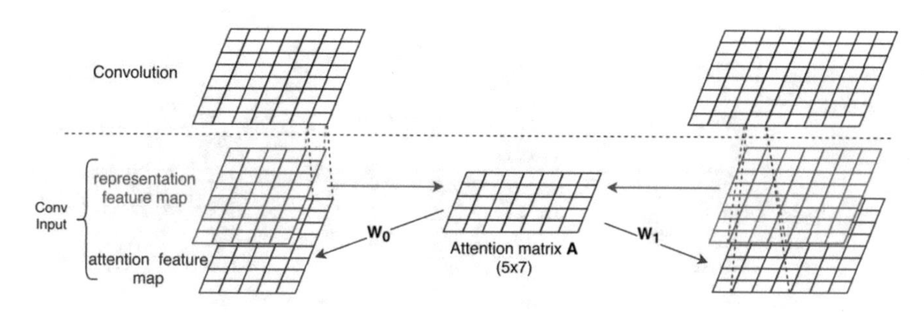

Fig. 5.11 The architecture of the ABCNN-1 model (Yin et al., 2016) ("©1963–2022 ACL, reprinted with permission")

and each column represents the attention weight of corresponding unit in the second sentence (s_1) regarding the first sentence (s_0). According to Fig. 5.11, two distinct attention feature maps (blue feature maps) are generated for each sequence and are passed to the convolution layer with the representation feature maps. Having the representation feature maps of the first ($\mathbf{F}_{0,r}[:, i]$) and second ($\mathbf{F}_{1,r}[:, i]$) sentences, the attention matrix $A \in R^{s \times s}$ and attention feature maps ($F_{0,a}$,$F_{1,a}$) are built as follows:

$$
\begin{aligned}
A_{i,j} &= match - score\left(F_{0,r}[:, i], F_{1,r}[:, j]\right) \\
match - score \ &= \frac{1}{(1 + |\mathrm{x} - \mathrm{y}|)}
\end{aligned}
\tag{5.106}
$$

where $W_0 \in R^{d \times s}$ and $W_1 \in R^{d \times s}$ are the model parameters. Representation feature map and attention feature map are jointly passed to the convolution layer with an order three tensor.

The architecture of the ABCNN-2 model is shown in Fig. 5.12. In this model, attention is applied to the output of convolution. Attention weight of each unit from s_0 and s_1 sentences, given the output of convolution layer $\mathbf{F}^c_{0,r} \in \mathbf{R}^{d \times (s_i + w - 1)}$ and $\mathbf{F}^c_{1,r} \in \mathbf{R}^{d \times (s_i + w - 1)}$ for s_0 and s_1 sentences, respectively, is computed as follows:

$$
\begin{aligned}
a_{0,j} &= \sum A[j, :] \\
a_{1,j} &= \sum A[:, j]
\end{aligned}
\tag{5.107}
$$

$$
\begin{aligned}
F^p_{0,r}[:, j] &= \sum_{k=j:j+w} a_{0,k} F^c_{0,r}[:, k], \quad j = 1 \ldots s_0 \\
F^p_{1,r}[:, j] &= \sum_{k=j:j+w} a_{1,k} F^c_{1,r}[:, k], \quad j = 1 \ldots s_1
\end{aligned}
\tag{5.108}
$$

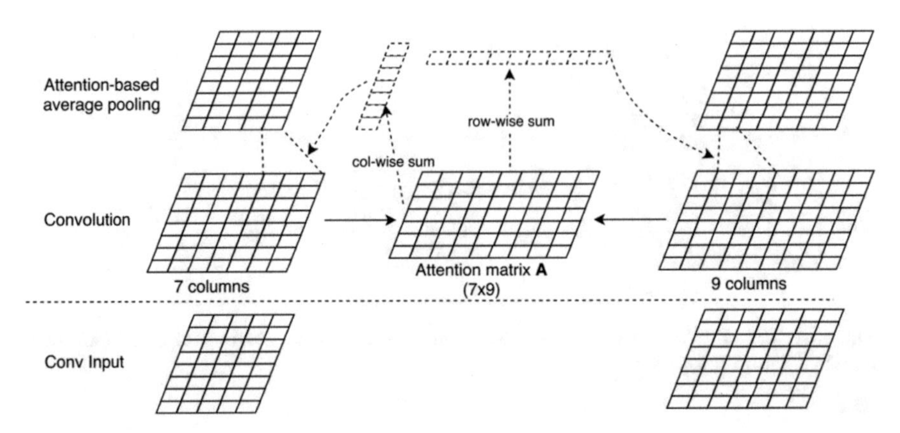

Fig. 5.12 The architecture of ABCNN-2 model (Yin et al., 2016) ("©1963–2022 ACL, reprinted with permission")

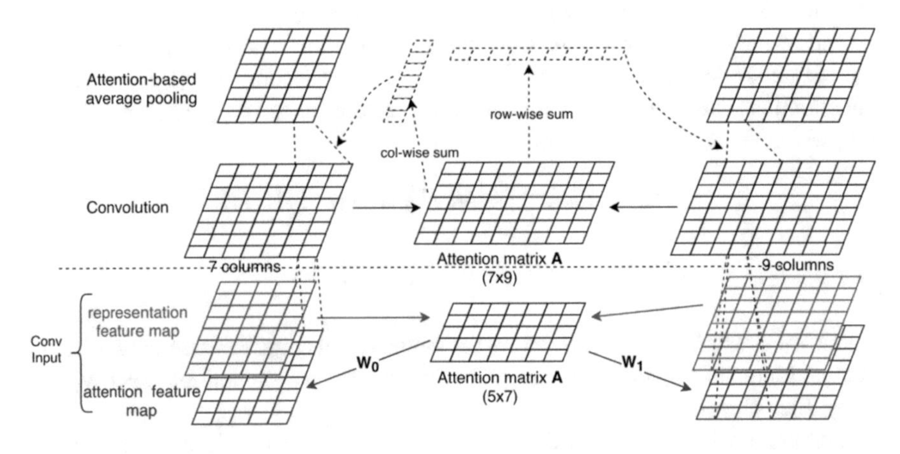

Fig. 5.13 The architecture of the ABCNN-3 model (Yin et al., 2016) ("©1963–2022 ACL, reprinted with permission")

where $a_{0,j}$ and $a_{1,j}$ are the attention weights of j-th unit in s_0 and s_1 sentences, respectively. $F^p_{1,r}[:, j]$ and $F^p_{0,r}[:, j]$ are the outputs of attentional pooling in ABCNN-2.

ABCNN-2 model has a smaller number of trainable parameters than ABCNN-1 model which results in lower chance of overfitting and faster training.

The architecture of the ABCNN-3 model is shown in Fig. 5.13. ABCNN-3 combines both ABCNN-2 and ABCNN-1 models by using attention after and before the convolution layer.

Creating a question-aware representation of the answer sentence is highly important as it is mainly influenced by informative words of the answer sentence which

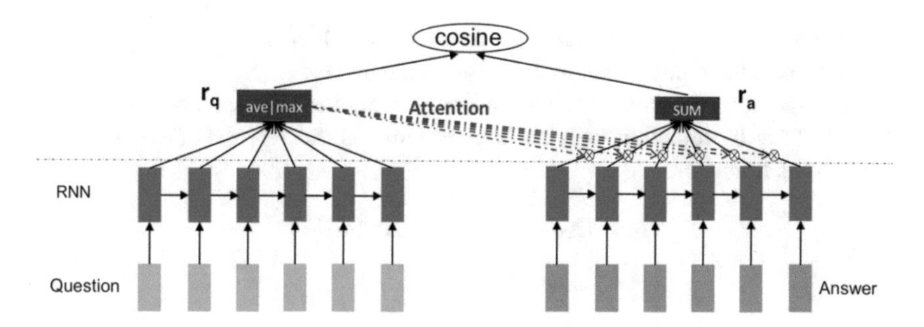

Fig. 5.14 The architecture of the OARNN model (Wang et al., 2016) ("©1963–2022 ACL, reprinted with permission")

help in answering the question. For example, consider the following candidate answer and questions:

A: *Michael Jordan abruptly retired from Chicago Bulls before the beginning of the 1993–1994 NBA season to pursue a career in baseball.*
Q1: *When did Michael Jordan retired from NBA?*
Q2: *Which sports does Michael Jordan participates after his retirement from NBA?*

For answering the first question, the representation generated for the answer sentence should mostly be influenced by "before the beginning of the 1993–1994 NBA season" as this phrase answers the question. While given the second question, the model should pay more attention to the phrase "pursue a career in baseball" in generating the representation of the answer sentence. This example clearly demonstrates the importance of the attention mechanism in building the representation of sentences in answer sentence selection activity.

OARNN (Outer Attention-based RNN) (Wang et al., 2016) is shown in Fig. 5.14. As can be seen, the attention mechanism is applied to output hidden states of RNN to obtain a question-aware representation for the answer sentence. The input of OARNN is question $Q = \{q_1, q_2, \cdots, q_n\}$ with length n and candidate answer $S = \{s_1, s_2, \cdots, s_m\}$ with length m. Embedding of the question sentence is obtained by using an embedding layer D and is given to RNN. The last hidden state of RNN or average of hidden states is considered as the representation of the question sentence. Representation of each hidden state is computed as follows:

$$\begin{aligned} \mathbf{X} &= \mathbf{D}\left[q_1, q_2, \ldots, q_n\right] \\ \mathbf{h}_t &= \sigma\left(\mathbf{W}_{ih}\mathbf{x}_t + \mathbf{W}_{hh}\mathbf{h}_{t-1} + \mathbf{b}_h\right) \\ \mathbf{y}_t &= \sigma\left(\mathbf{W}_{ho}\mathbf{h}_t + \mathbf{b}_o\right) \end{aligned} \tag{5.109}$$

where W_{ih}, W_{ho}, and W_{hh} are the parameters to be learned, b_o and b_h are the biases, σ is the activation function, and D converts the input word to its embedding in

R^d space. The last hidden state (h_n) or average of hidden states ($\frac{1}{n}\sum_{t=1}^{n} h_t$) is considered as the representation of question sentence r_q.

For generating the answer representation r_a, weighted representation of answer's hidden states by attention weights is considered as the representation of answer sentence as follows:

$$\begin{aligned}
&\mathbf{H}_a = [\mathbf{h}_a(1), \mathbf{h}_a(2), \ldots, \mathbf{h}_a(m)] \\
&s_t \propto f_{attention}\left(\mathbf{r}_q, \mathbf{h}_a(t)\right) \\
&\tilde{\mathbf{h}}_a(t) = \mathbf{h}_a(t) s_t \\
&\mathbf{m}(t) = \tanh\left(\mathbf{W}_{hm}\mathbf{h}_a(t) + \mathbf{W}_{qm}\mathbf{r}_q\right) \\
&f_{attention}\left(\mathbf{r}_q, \mathbf{h}_a(t)\right) = \exp\left(\mathbf{w}_{ms}^T \mathbf{m}(t)\right) \\
&\mathbf{r}_a = \sum_{t=1}^{m} \tilde{\mathbf{h}}_a(t)
\end{aligned} \tag{5.110}$$

where $h_a(t)$ represents the output hidden states at time step t, s_t is the attention weight generated for time step t, W_{hm} and W_{qm} are the attention matrices which are model parameters to be learned, and w_{ms} is the attentive weight vector.

The last hidden states in RNNs carry more information about the sentence as they know about all of the words within the sentence. As a result, the attention weight assigned to the hidden states increases, moving toward the last hidden states. This is called the attention bias problem. For mitigating the attention bias problem, three other Inner Attention-based RNNs (IARNN) are proposed.

IARNN-WORD applies the attention mechanism before RNN, and answer representation is generated by average pooling over all output hidden states. The architecture of IARNN-WORD model is shown in Fig. 5.15. GRU is selected for modeling the answer representation rather than LSTM as it has less parameters. The

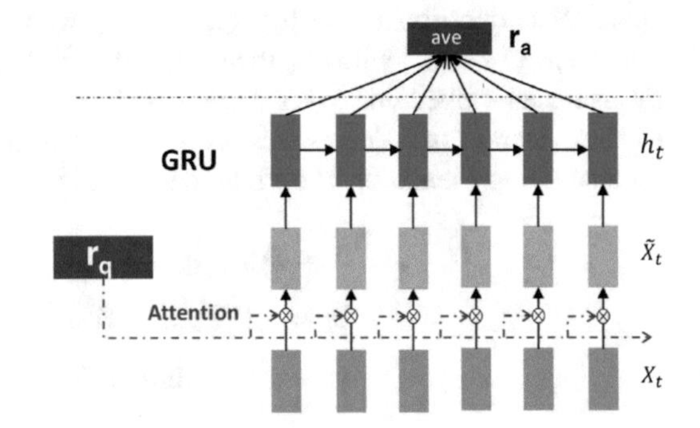

Fig. 5.15 The architecture of the IARNN-WORD model (Wang et al., 2016) ("©1963–2022 ACL, reprinted with permission)

weighted representation of answer words is passed to GRU as follows:

$$
\begin{aligned}
\alpha_t &= \sigma\left(\mathbf{r}_q^T \mathbf{M}_q \mathbf{x}_t\right) \\
\tilde{\mathbf{x}}_t &= \alpha_t * \mathbf{x}_t \\
\mathbf{z}_t &= \sigma\left(\mathbf{W}_{xz}\tilde{\mathbf{x}}_t + \mathbf{W}_{hz}\mathbf{h}_{t-1}\right) \\
\mathbf{f}_t &= \sigma\left(\mathbf{W}_{xf}\tilde{\mathbf{x}}_t + \mathbf{W}_{hf}\mathbf{h}_{t-1}\right) \\
\tilde{\mathbf{h}}_t &= \tanh\left(\mathbf{W}_{xh}\tilde{\mathbf{x}}_t + \mathbf{W}_{hh}\left(\mathbf{f}_t \odot \mathbf{h}_{t-1}\right)\right) \\
\mathbf{h}_t &= (1 - \mathbf{z}_t) \odot \mathbf{h}_{t-1} + \mathbf{z}_t \odot \tilde{\mathbf{h}}_t
\end{aligned}
\tag{5.111}
$$

where $\tilde{x}_t$ is the weighted representation of answer token x_t and W_{xz}, W_{hz}, W_{xf}, W_{hh}, and W_{xh} are the model parameters.

In the IARNN-WORD model, attention is applied to word level, and contextual information is missed in attending answer words. In some cases, contextual meaning of a word in the answer sentence is related to the question, while the token without considering the context is not informative and helpful. Word-level attention cannot detect the phrase-level or contextual relation of answer words. IARNN-CONTEXT model is proposed for mitigating the mentioned problem.

IARNN-CONTEXT uses contextual information of the answer sentence in calculating the attention weight of answer tokens. The architecture of the IARNN-CONTEXT model is shown in Fig. 5.16. Given the contextual information, attentional representation of the answer tokens $\tilde{x}_t$ is calculated as follows:

$$
\begin{aligned}
\mathbf{w}_C(t) &= \mathbf{M}_{hc}\mathbf{h}_{t-1} + \mathbf{M}_{qc}\mathbf{r}_q \\
\alpha_C^t &= \sigma\left(\mathbf{w}_C^T(t)\mathbf{x}_t\right) \\
\tilde{\mathbf{x}}_t &= \alpha_C^t * \mathbf{x}_t
\end{aligned}
\tag{5.112}
$$

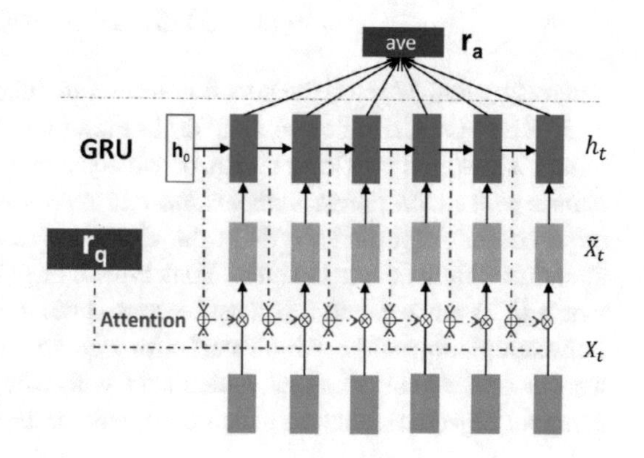

Fig. 5.16 The architecture of the IARNN-CONTEXT model (Wang et al., 2016) ("©1963–2022 ACL, reprinted with permission")

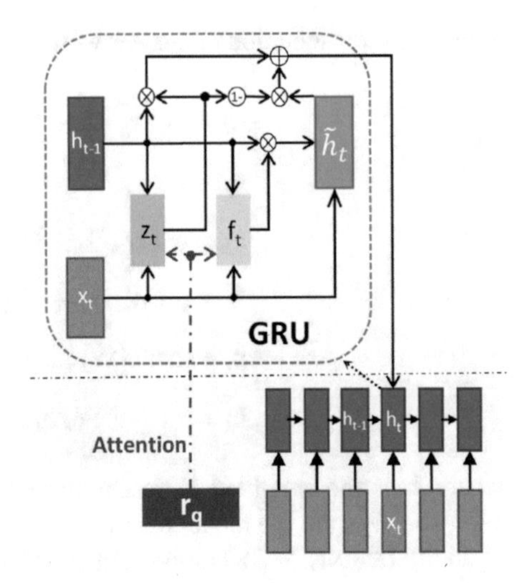

Fig. 5.17 The architecture of IABRNN-GATE (Wang et al., 2016) ("©1963–2022 ACL, reprinted with permission")

where the output of previous hidden state h_{t-1} is used as context information and M_{hc} and M_{qc} are the attention weight matrices to be learned.

In RNNs, the information passed to further hidden states or flow of information is controlled by update and forget gates. The amount of information passed to the next hidden states could be controlled by applying attention to the mentioned active gates.

IARNN-GATE works by incorporating the attention weights into RNN gates. The architecture of the IARNN-GATE model is shown in Fig. 5.17. The GRU gates are updated as follows:

$$\begin{aligned} z_t &= \sigma\left(W_{xz}x_t + W_{hz}h_{t-1} + M_{qz}r_q\right) \\ f_t &= \sigma\left(W_{xf}x_t + W_{hf}h_{t-1} + M_{qf}r_q\right) \\ \tilde{h}_t &= \tanh\left(W_{xh}x_t + W_{hh}\left(f_t \odot h_{t-1}\right)\right) \\ h_t &= (1 - z_t) \odot h_{t-1} + z_t \odot \tilde{h}_t \end{aligned} \tag{5.113}$$

where M_{qz} and M_{qf} are the attention weight matrices to be learned.

IARNN-OCCAM controls and limits the summation of the attention words. This model is inspired by Occam's Razor which says: *Among the whole words set, we choose those with fewest number that can represent the sentence.* The number of required and important words in the answer sentence could be controlled by the attention weight devoted to them. Each type of question requires various numbers of words in its answer, and this must be considered in punching the attention weights. For example, questions which start with *who* and *when* need fewer words in their answer rather than questions which start with *why* or *how*. This model works by using an objective function which considers the type of question in specifying the limit of attention weights as follows:

$$
\begin{aligned}
n_p^i &= \max\left\{\mathbf{w}_{qp}^T \mathbf{r}_q^i, \lambda_q\right\} \\
J_i^* &= J_i + n_p^i \sum_{t=1}^{mc} \alpha_t^i
\end{aligned}
\tag{5.114}
$$

where λ_q is a small positive parameter which must be tuned, J is the original objective, r_q^i is the representation of question Q^i, and α_t^i is the attention weight which is used in the IARNN-WORD and IARNN-CONTEXT models.

We have presented more than ten hybrid neural models for answer sentence selection. The hybrid architectures mostly use the interaction between two input sentences as attention weights for generating the sentence representations. In other words, they follow a compare-aggregate architecture. After generating the sentence representations, they are matched to predict the matching score. For example, BiMPM, Comp-Clip, and the model proposed by (Bian et al., 2017) have a compare-aggregate architecture. The attention mechanism is considered as an interaction component in hybrid models. Different types of attention mechanisms are proposed and used in these models. We have discussed the benefits and the limitations of each attention mechanism in this section. Both CNN and RNNs are widely used in the hybrid models for generating the context representation of the question and the answer sentences or aggregating them.

5.4 Summary

In Sect. 5.2, we explained traditional models in TextQA that are not based on the neural networks. The neural network-based models are explained in Sect. 5.3 in three different categories, namely, representation-based models (Sect. 5.3.1), interaction-based models (Sect. 5.3.2), and hybrid models (Sect. 5.3.3). As can be seen, a large number of neural-based models in this area belong to the hybrid category.

References

Bian, W., Li, S., Yang, Z., Chen, G., & Lin, Z. (2017). A compare-aggregate model with dynamic-clip attention for answer selection. In *Proceedings of the 2017 ACM on Conference on Information and Knowledge Management, CIKM '17* (pp. 1987–1990). ACM. ISBN 978-1-4503-4918-5. https://doi.org/10.1145/3132847.3133089

Devlin, J., Chang, M. W., Lee, K., & Toutanova, K. (2019). Bert: Pre-training of deep bidirectional transformers for language understanding. In *NAACL-HLT*.

Garg, S., Vu, T., & Moschitti, A. (2020). Tanda: Transfer and adapt pre-trained transformer models for answer sentence selection. In *Thirty-Fourth AAAI Conference on Artificial Intelligence*
He, H., & Lin, J. (2016). Pairwise word interaction modeling with deep neural networks for semantic similarity measurement. In *Proceedings of the 2016 Conference of the North American Chapter of the Association for Computational Linguistics: Human Language Technologies*, San Diego, California (pp. 937–948). Association for Computational Linguistics. https://doi.org/10.18653/v1/N16-1108
Hochreiter, S., & Schmidhuber, J. (1997). Long short-term memory. *Neural Computation, 9*(8), 1735–1780.
Kwiatkowski, T., Palomaki, J., Redfield, O., Collins, M., Parikh, A., Alberti, C., Epstein, D., Polosukhin, I., Devlin, J., Lee, K., Toutanova, K., Jones, L., Kelcey, M., Chang, M.-W., Dai, A., Uszkoreit, J., Le, Q., & Petrov, S. (2019). Natural questions: A benchmark for question answering research. *Transactions of the Association for Computational Linguistics, 7*, 453–466.
Lai, T., Tran, Q. H., Bui, T., & Kihara, D. (2019). A gated self-attention memory network for answer selection. In *Proceedings of the 2019 Conference on Empirical Methods in Natural Language Processing and the 9th International Joint Conference on Natural Language Processing (EMNLP-IJCNLP)*, Hong Kong, China (pp. 5953–5959). Association for Computational Linguistics. https://doi.org/10.18653/v1/D19-1610
Laskar, M. T. R., Huang, X., & Hoque, E. (2020). Contextualized embeddings based transformer encoder for sentence similarity modeling in answer selection task. In *Proceedings of the 12th Language Resources and Evaluation Conference*, Marseille, France (pp. 5505–5514). European Language Resources Association.
Liu, Y., Ott, M., Goyal, N., Du, J., Joshi, M., Chen, D., Levy, O., Lewis, M., Zettlemoyer, L., & Stoyanov, V. (2019). Roberta: A robustly optimized bert pretraining approach. arXiv preprint arXiv:1907.11692.
Mikolov, T., Sutskever, I., Chen, K., Corrado, G. S., & Dean, J. (2013). Distributed representations of words and phrases and their compositionality. In *Proceedings of the 26th International Conference on Neural Information Processing Systems - Volume 2, NIPS'13* (pp. 3111–3119). Curran Associates Inc.
Momtazi, S., & Klakow, D. (2009). A word clustering approach for language model-based sentence retrieval in question answering systems. In *Proceedings of the Annual International ACM Conference on Information and Knowledge Management (CIKM)* (pp. 1911–1914). ACM.
Momtazi, S., & Klakow, D. (2011). Trained trigger language model for sentence retrieval in QA: Bridging the vocabulary gap. In *Proceedings of the Annual International ACM Conference on Information and Knowledge Management (CIKM)*.
Momtazi, S., & Klakow, D. (2015). Bridging the vocabulary gap between questions and answer sentences. *Information Processing and Management, 51*(5), 595–615. ISSN 0306-4573. https://doi.org/10.1016/j.ipm.2015.04.005
Murdock, V., Croft, W. B. (2004). Simple translation models for sentence retrieval in factoid question answering. In *SIGIR 2004*.
Pennington, J., Socher, R., & Manning, C. D. (2014). Glove: Global vectors for word representation. In *Proceedings of the 2014 Conference on Empirical Methods in Natural Language Processing (EMNLP)*, Doha, Qatar (pp. 1532–1543). Association for Computational Linguistics. https://doi.org/10.3115/v1/D14-1162
Peters, M., Neumann, M., Iyyer, M., Gardner, M., Clark, C., Lee, K., & Zettlemoyer, L. (2018). Deep contextualized word representations. In *Proceedings of the 2018 Conference of the North American Chapter of the Association for Computational Linguistics: Human Language Technologies, Volume 1 (Long Papers)*, New Orleans, Louisiana (pp. 2227–2237). Association for Computational Linguistics. https://doi.org/10.18653/v1/N18-1202
Severyn, A., & Moschitti, A. (2015). Learning to rank short text pairs with convolutional deep neural networks. In *SIGIR*
Tan, M., Dos Santos, C., Xiang, B., & Zhou, B. (2016). Improved representation learning for question answer matching. In *Proceedings of the 54th Annual Meeting of the Association*

for Computational Linguistics (Volume 1: Long Papers), Berlin, Germany (pp. 464–473). Association for Computational Linguistics. https://doi.org/10.18653/v1/P16-1044

Tay, Y., Phan, M. C., Tuan, L. A., & Hui, S. C. (2017). Learning to rank question answer pairs with holographic dual lstm architecture. In *Proceedings of the 40th International ACM SIGIR Conference on Research and Development in Information Retrieval*, SIGIR '17 (pp. 695–704), New York, NY, USA. ACM. ISBN 978-1-4503-5022-8. https://doi.org/10.1145/3077136.3080790

Tay, Y., Tuan, L. A., & Hui, S. C. (2018). Multi-cast attention networks. In *Proceedings of the 24th ACM SIGKDD International Conference on Knowledge Discovery & Data Mining, KDD '18*, New York, NY, USA (pp. 2299–2308). Association for Computing Machinery.

Wan, S., Lan, Y., Guo, J., Xu, J., Pang, L., & Cheng, X. (2016a) A deep architecture for semantic matching with multiple positional sentence representations. In *Proceedings of the Thirtieth AAAI Conference on Artificial Intelligence, AAAI'16* (pp. 2835–2841). AAAI Press.

Wan, S., Lan, Y., Xu, J., Guo, J., Pang, L., & Cheng, X. (2016b). Match-SRNN: Modeling the recursive matching structure with spatial RNN. In *IJCAI*.

Wang, A., Singh, A., Michael, J., Hill, F., Levy, O., & Bowman, S. R. (2018). GLUE: A multi-task benchmark and analysis platform for natural language understanding. In *Proceedings of the 2018 EMNLP Workshop BlackboxNLP: Analyzing and Interpreting Neural Networks for NLP*, Brussels, Belgium (pp. 353–355). Association for Computational Linguistics. https://doi.org/10.18653/v1/W18-5446

Wang, B., Liu, K., & Zhao, J. (2016). Inner attention based recurrent neural networks for answer selection. In *Proceedings of the 54th Annual Meeting of the Association for Computational Linguistics (Volume 1: Long Papers)*, Berlin, Germany (pp. 1288–1297). Association for Computational Linguistics. https://doi.org/10.18653/v1/P16-1122

Wang, D., & Nyberg, E. (2015). A long short-term memory model for answer sentence selection in question answering. In *Proceedings of the 53rd Annual Meeting of the Association for Computational Linguistics and the 7th International Joint Conference on Natural Language Processing (Volume 2: Short Papers)*, Beijing, China (pp. 707–712). Association for Computational Linguistics. https://doi.org/10.3115/v1/P15-2116

Wang, S., & Jiang, J. (2017). A compare-aggregate model for matching text sequences. In *Proceedings of the 5th International Conference on Learning Representations*. International Conference on Learning Representations (ICLR).

Wang, Z., Hamza, W., & Florian, R. (2017). Bilateral multi-perspective matching for natural language sentences. In *Proceedings of the Twenty-Sixth International Joint Conference on Artificial Intelligence, IJCAI-17* (pp. 4144–4150). https://doi.org/10.24963/ijcai.2017/579

Yang, R., Zhang, J., Gao, X., Ji, F., & Chen, H. (2019). Simple and effective text matching with richer alignment features. In *Proceedings of the 57th Annual Meeting of the Association for Computational Linguistics* (pp. 4699–4709). Association for Computational Linguistics.

Yin, W., Schütze, H., Xiang, B., & Zhou, B. (2016). Abcnn: Attention-based convolutional neural network for modeling sentence pairs. *Transactions of the Association for Computational Linguistics, 4*, 259–272.

Yoon, S., Dernoncourt, F., Kim, D. S., Bui, T., & Jung, K. (2019). A compare-aggregate model with latent clustering for answer selection. In *Proceedings of the 28th ACM International Conference on Information and Knowledge Management* (pp. 2093–2096).

Yu, L., Hermann, K. M., Blunsom, P., & Pulman, S. (2014). Deep learning for answer sentence selection. In *Deep Learning and Representation Learning Workshop: NIPS 2014*. abs/1412.1632, 2014.

Chapter 6
Question Answering over Knowledge Base

Abstract This chapter describes the available research studies in KBQA from two perspectives, including traditional models and deep learning-based models. In each category, various researches will be presented by discussing their architecture. This chapter includes a comprehensive comparison of proposed methods and describes state-of-the-art models in terms of both simple and complex QA.

6.1 Introduction

KBQA is defined as extracting the answer entity from the knowledge base given an input question. Information is stored in the form of triple facts in a knowledge base where each fact indicates a relation between two entities. Each triple includes a head entity which is connected to a tail entity by a relation.

To answer some questions, only one fact from the knowledge base is required, while answering some questions requires reasoning over multiple facts. Depending on the type of the questions, which can be simple or complex, the available techniques for KBQA are divided into two main categories.

The traditional methods for KBQA are explained in Sect. 6.2; the methods proposed for simple questions are discussed in Sect. 6.3; the models proposed for complex questions are introduced in Sect. 6.4.

6.2 Traditional Models

The traditional models proposed for QA over KB do not benefit from neural architectures. We will discuss three traditional models in this section.

Unger et al. (2012) proposed a model which aims to find a mapping between language processing and SPARQL queries. Most QA systems map questions to a triple representation to produce SPARQL query templates. These approaches are not able to produce complex queries. In this model, an approach is used that uses a

S. Momtazi, Z. Abbasiantaeb, *Question Answering over Text and Knowledge Base*,
https://doi.org/10.1007/978-3-031-16552-8_6

natural language parser and produces the SPARQL query template using the parse of the question. This approach is done through three parts:

- **Template generation:** To convert a natural language question into a SPARQL template, the Pythia system, which is a natural language parser, is adopted. Therefore, the semantic representation is used to generate the SPARQL templates. This part contains domain-dependent expressions, and before mapping URIs, they are not specified. Instead, the templates contain slots which are built based on POS information provided by the Stanford POS tagger. These are some rules for this step:
  - Named entities are considered as resources. Thus, a lexical entry is built containing a resource slot.
  - Nouns can be referring to classes or properties. Thus, two entries are built such that one of them contains a class slot and another one contains a property slot.
  - Verbs mostly refer to properties. Thus, an entry is built with a property slot. But the verb does not always contribute to the query structure; thus, another entry is built without any slots corresponding to the verb.
- **Entity identification:** After the first part, there are several slots that must be replaced by appropriate URIs. This part is done using entity and property detection methods based on similarity.
- **Query ranking:** In the final step, there are several possible SPARQL queries. For each slot in each query, two scores are computed: a similarity score that is computed based on similarity between the entity and its synsets (synonyms) from WordNet and a prominence score.

Yao and Van Durme (2014) proposed a model that is inspired by the way humans search a knowledge base to find answers to a question. The model works in two steps: (1) constructs a topic graph by extracting a view of the knowledge base that is relevant to the given question and (2) extracts the discriminative features from the question and the topic graph and uses them for the answer extraction process. The workflow of the model is described in the following.

- **Topic graph construction:** Consider we are given the question "What is the capital of France?" to find its answer from Freebase. To answer the mentioned question, several points are taken into the consideration:

  1. The two dependency relations nsubj(what, capital) and prep_of(capital, France) demonstrate that the question asks about a capital,
  2. According to the relation prep_of(capital, France), the capital refers to a France,
  3. As France refers to a county and we have the relation nn(capital, france), the capital refers to the name of a city because the capital of a country is a city.

  The above points reflect the importance of dependency-based features. In addition, some linguistic information, which is listed below, is useful.

1. Question words (qword) including who, how, where, why, whom, which, when, what, and whose
2. Question focus (qfocus) that indicates the EAT of the question like money, place, time, and person. The noun dependent on the question word is considered as question focus.
3. Question verb (qverb).
4. Question topic (qtopic) is used for retrieving the relevant Freebase pages. Named entities of the question are considered as the question topic.

The dependency parse tree is converted into a question graph in the following three steps.

1. Replace the nodes including question features with the corresponding question feature (what → qword=what).
2. Replace the named entity nodes with the type of the named entity (France → qtopic=county).
3. Eliminate the leaf nodes that are determiner, preposition, or punctuation.

An example of the dependency features with its converted graph (question feature graph) for the question "What is the capital of France?" is shown in the left and the right parts of Fig. 6.1, respectively. The question feature graph is used for extracting the features. Each edge $e(Standing)$ connecting the two source s and target t nodes is transformed to features s, t, $s|t$ and $s|e|t$.

The topic graph is constructed by selecting the nodes in a few hop distance of the topic node in Freebase. The properties are also highly important in recognizing the answer. The properties connect an entity to a string for explaining the features of the entity node. The argument of each property is connected to only one entity node, while both arguments of the relations are entity nodes that can be connected to several other entities.

The corresponding topic graph of our example is represented in Fig. 6.2. As can be seen, some properties such as type connected to "France" and "Paris" nodes help us to identify the capital of *France*.

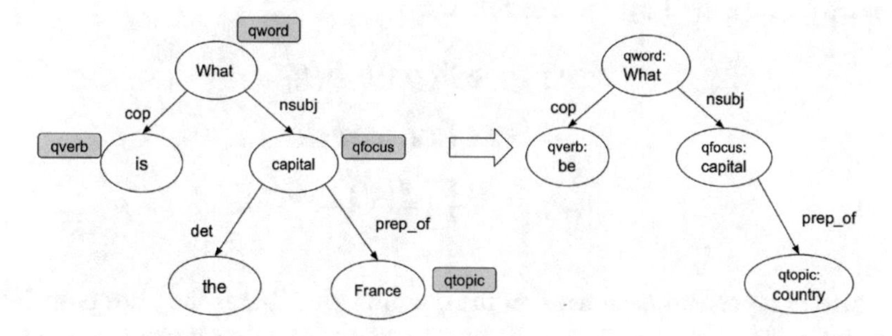

Fig. 6.1 An example of a dependency parse tree and corresponding feature graph (Yao & Van Durme, 2014)

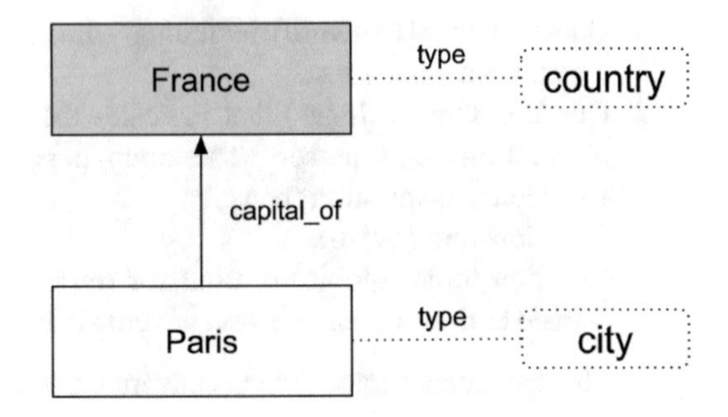

Fig. 6.2 The corresponding topic graph of the mentioned example which is extracted from Freebase (Yao & Van Durme, 2014). The properties are represented by dash lines, and the entities are depicted by solid lines

Freebase relations can be mapped to questions by lexical overlap or paraphrasing. An alignment model can be developed to estimate the probability of mapping a question to a relation. The probability of each relation given a question is calculated, and the ranked list of relations is used as a feature.

The question features and Freebase features for each node are combined by pair-wise concatenation. This helps us to model the association of answer nodes and the question pattern. In the case of utilizing a log-linear model, a feature like qfocus=money|node_type=currency must be given a higher weight than a feature like qfocus=money|node_type=person.

- **Relation mapping:** The goal of this task is estimating a probability for each relation given a question to predict the most likely relation for the given question. Consider the following questions along with their most probable relation:

Q: *Who is the father of King George VI?*
Relation: *people.person.parents*

Q: *Who is the father of the Periodic Table?*
Relation: *law.invention.inventor*

The above examples reflect the fact that all of the question words influence the choice of relation. As a result, the model must consider all of the question words in calculating the probability. Thus, the conditional probability is calculated based on the naïve Bayes rule as follows:

$$\begin{aligned} \tilde{P}(R \mid Q) &\propto \tilde{P}(Q \mid R)\tilde{P}(R) \\ &\approx \tilde{P}(\mathbf{w} \mid R)\tilde{P}(R) \\ &\approx \prod_{w} \tilde{P}(w \mid R)\tilde{P}(R) \end{aligned} \tag{6.1}$$

where w represents each word of the question Q, $\tilde{P}(R)$ is the prior probability of the relation R, and $\tilde{P}(w \mid R)$ is the probability of each word w given the relation R.

In another case, we consider the relation R as a sequence of sub-relations r_i. For example, the relation people.person.parents is the concatenation of the sub-relations people, person, and parents. We compute the backoff probability of the relation R given the question Q as follows:

$$\begin{aligned}\tilde{P}_{\text{backoff}}(R \mid Q) &\approx \tilde{P}(\mathbf{r} \mid Q) \\ &\approx \prod_r \tilde{P}(r \mid Q) \\ &\propto \prod_r \tilde{P}(Q \mid r)\tilde{P}(r) \\ &\approx \prod_r \prod_w \tilde{P}(w \mid r)\tilde{P}(r)\end{aligned} \tag{6.2}$$

The value of the backoff probability is much less than the original probability due to the two sets of multiplications. To keep it with the same scale as the original probability, we normalize it with the sub-relations count.

The ClueWeb09 dataset[1] and FACC1 (Orr et al., 2011) are used for estimating the probabilities. FACC1 indicates the index of Freebase entities in English segments of the first version of ClueWeb. FACC1 contains 340 documents that are annotated with 1 entity, and the whole collection is annotated with an average of 15 entity annotations per document.

The binary relations within Freebase are associated with a collection of sentences that each sentence includes one argument of the relation. Then, the association of words with the relation is learned to estimate the probabilities $\tilde{P}(w \mid R)$ and $\tilde{P}(w \mid r)$. The probabilities $\tilde{P}(R)$ and $\tilde{P}(r)$ are estimated by counting the number of times that the relation R is annotated in the dataset. The process of learning the above probabilities is performed in the following five steps.

1. The HTML document is divided into sentences by using the NLTK tool (Bird & Loper, 2004), and the sentences including the Freebase entities, which are connected to each other with at least one relation in Freebase, are selected.
2. The output of the previous step constructs two parallel corpora for indicating "relation-sentence" and "sub-relation-sentence" associations. The "relation-sentence" association is used for calculating the probabilities $\tilde{P}(R)$ and $\tilde{P}(w \mid R)$, and "sub-relation-sentence" association is utilized for calculating $\tilde{P}(r)$ and $\tilde{P}(w \mid r)$. The IBM alignment model (Brown et al., 1993) is utilized for estimating the translation probability by GIZA++ (Och & Ney, 2003) because one side of the parallel corpora is not a natural language sentence.
3. A co-occurrence matrix is generated by aligning the relations and words from sentences.
4. By using the co-occurrence matrix, the probabilities $\tilde{P}(w \mid R)$, $\tilde{P}(w \mid r)$, $\tilde{P}(R)$, and $\tilde{P}(r)$ are calculated.

[1] http://lemurproject.org/clueweb09/.

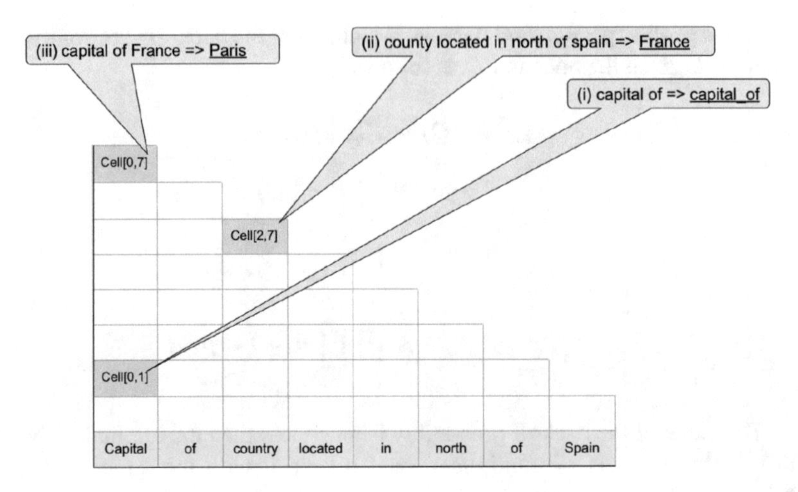

Fig. 6.3 Illustration of an example in the translation-based model proposed by Bao et al. (2014)

Bao et al. (2014) proposed a unified framework based on machine translation. The given question is parsed using CYK parsing, and each cell is translated to the answers of its span like machine translation. A question translation technique based on semantic parsing is utilized to generate the translation of each cell using the question pattern and relation expressions. The answer can be extracted from the root cell. A linear model is used for modeling the derivations produced in the translation process, and feature weights are tuned by using a minimum error rate training (Och, 2003).

An example of answering the question "Capital of country located in north of Spain" is represented in Fig. 6.3. The question is translated to its answer in three steps: (1) translating the capital to *capital of* relation, (2) translating the phrase country located in north of Spain to an answer *France*, and (3) translating both the *capital of* and *France* to *Paris*.

A set of triple and answer pairs $\langle \mathcal{D}, \mathcal{A} \rangle$ are generated for the input question Q from knowledge base $\mathcal{KB}$ and ranked according to the following probability:

$$P(\langle \mathcal{D}, \mathcal{A} \rangle \mid \mathcal{KB}, Q) = \frac{\exp\left\{\sum_{i=1}^{M} \lambda_i \cdot h_i(\langle \mathcal{D}, \mathcal{A} \rangle, \mathcal{KB}, Q)\right\}}{\sum_{\langle \mathcal{D}', \mathcal{A}' \rangle \in \mathcal{H}(Q)} \exp\left\{\sum_{i=1}^{M} \lambda_i \cdot h_i\left(\langle \mathcal{D}', \mathcal{A}' \rangle, \mathcal{KB}, Q\right)\right\}} \tag{6.3}$$

where each assertion (fact) is stored in $\mathcal{KB}$ in the form of [subject_entity, predicate, object_entity] ($\left\{e_{sbj}^{ID}, p, e_{obj}^{ID}\right\}$), $\mathcal{H}(Q)$ indicates the search space, $\mathcal{D}$ includes a set of ordered triples $\{t_1, \cdots, t_n\}$ that each triple beginning i and end j indexes of the question span are indicated within each triple as $t = \left\{e_{sbj}, p, e_{obj}\right\}_i^j \in \mathcal{D}$, $h_i(\cdot)$

indicates the feature function, and λ_i represents the associated feature weight of $h_i(\cdot)$.

The list of triples is sorted according to the order of translation steps for translating question Q to answer $\mathcal{A}$.

The model includes four main steps to compute the above probability that are explained in the following.

- **Search space generation:** The set $\mathcal{H}(Q)$ is created by using question translation method, which is proposed in this chapter, according to Algorithm 3. In the first part of this algorithm, the question translation method translates each unary span Q_i^j and forms a formal triple set T given the input knowledge graph as $QTrans(Q_i^j, \mathcal{KB})$. The subject entity of the each retrieved triple exists in the corresponding unary span of the question sentence. The predicate of the each retrieved triple is the meaning represented by the subject entity in unary span of the question. The second part of the algorithm forms the bigger spans and updates their triples by calling the $QTrans$ function.

Algorithm 3 Translation-based method for KBQA (Bao et al., 2014)

for $l = 1$ to $|Q|$ **do**
  for all i, j s.t. $j - i = l$ **do**
    $\mathcal{H}(Q_i^j) = \emptyset$
    $T = \text{QTrans}(Q_i^j, \mathcal{KB})$
    for each *formal triple* $t \in T$ **do**
      create a new derivation d
      $d.\mathcal{A} = t.e_{obj}$
      $d.\mathcal{D} = \{t\}$
      update the model score of d
      insert d to $\mathcal{H}(Q_i^j)$
    end for
  end for
end for
for $l = 1$ to $|Q|$ **do**
  for all i, j s.t. $j - i = l$ **do**
    for all m s.t. $i \leq m < j$ **do**
      for $d_l \in \mathcal{H}(Q_i^m)$ and $d_r \in \mathcal{H}\left(Q_{m+1}^j\right)$ **do**
        $Q_{\text{update}} = d_l.\mathcal{A} + d_r.\mathcal{A}$
        $T = Q\ \text{Trans}\left(Q_{\text{update}}, \mathcal{KB}\right)$
        for each *formal triple* $t \in T$ **do**
          create a new derivation d
          $d.\mathcal{A} = t.e_{obj}$
          $d \cdot \mathcal{D} = d_l \cdot \mathcal{D} \bigcup d_r.\mathcal{D} \bigcup \{t\}$
          update the model score of d
          insert d to $\mathcal{H}(Q_i^j)$
        end for
      end for
    end for
  end for
end for

- **Question translation:** The goal of the question translation algorithm is to translate each input question to a set of formal triples. The translation is performed in a way that (1) the subject entity of the triple occurs in the question, (2) the predicate is inferred based on the context of the subject entity in the question span, and (3) the object entity is considered as the answer of the question. The question translation method considers the input span as a single-relation question, and the responsibility of handling the multiple-relation questions is given to CYK parsing-based translation process. Two different question translation models, namely, question pattern-based and relation expression-based methods, are proposed.
- **Feature design:** Different feature sets $h_i(.)$ are proposed for using in the probability function 6.3 that are listed below.
  - $h_{question\_word(.)}$ emphasizes on including all of the question words. This function returns the count of co-occurred words in both the question and answer $\mathcal{A}$.
  - $h_{span(.)}$ checks the granularity of the spans and is defined as the number of spans in Q that are translated to formal triples.
  - $h_{syntax\_subtree(.)}$ is the number of the question spans which (1) are translated to triples with non-*Null* predicate and (2) are simultaneously covered by dependency sub-trees.
  - $h_{syntax\_constraint(.)}$ is the count of triples from $\mathcal{D}$, which are translated to the sub-questions produced by question decomposition in relation expression-based translation model.
  - $h_{triple(.)}$ is defined as the count of triples from $\mathcal{D}$ with non-*Null* predicate.
  - $h_{triple_{weight}}(.)$ is defined as the summation of the scores of triples as $\sum_{t_i \in \mathcal{D}} t_i$.
  - $h_{QP_{count}}(.)$ is the number of triples generated by a question pattern-based translation model.
  - $h_{RE_{count}}(.)$ is the number of triples generated by relation expression-based translation model.
  - $h_{staticrank_{sbj}}(.)$ is defined as the summation of static rank scores of all subject entities as $\sum_{t_i \in \mathcal{D}} t_i.e_{sbj}.static\_rank$.
  - $h_{staticrank_{obj}}(.)$ is defined as the summation of static rank scores of all object entities as $\sum_{t_i \in \mathcal{D}} t_i.e_{obj}.static\_rank$.
  - $h_{confidence_{obj}}(.)$ is defined as the summation of confidence scores of all object entities as $\sum_{t_i \in \mathcal{D}} t_i.e_{sbj}.static\_rank$.
- **Feature weight tuning:** Minimum error rate training algorithm (Och, 2003) is used for tuning the λ weights. The training objective is defined as minimizing the accumulated errors using the top 1 answers of the questions as follows:

$$\hat{\lambda}_1^M = \underset{\lambda_1^M}{\operatorname{argmin}} \sum_{i=1}^{N} \operatorname{Err}\left(\mathcal{A}_i^{ref}, \hat{\mathcal{A}}_i; \lambda_1^M\right)$$
$$\operatorname{Err}\left(\mathcal{A}_i^{\text{ref}}, \hat{\mathcal{A}}_i; \lambda_1^M\right) = 1 - \delta\left(\mathcal{A}_i^{\text{ref}}, \hat{\mathcal{A}}_i\right) \tag{6.4}$$

where N indicates the total count of questions, $\hat{\mathcal{A}}_i$ represents the top 1 answer candidate according to the feature weights λ_1^M, $\mathcal{A}_i^{\text{ref}}$ represents the correct answers as references of the i-th question, and δ function returns one when $\mathcal{A}_i^{\text{ref}}$ exist in $\hat{\mathcal{A}}_i$ and zero otherwise.

We have discussed three traditional models for QA over KB. The model proposed by Unger et al. (2012) directly converts the input question into a SPARQL query. The approach proposed by Yao and Van Durme (2014) is inspired based on human approach for answering the questions from KB. The approach first retrieves the question topics from KB and then investigates the relations of topic nodes to find the answer. The approach proposed by Bao et al. (2014) works based on translation theory. It translates each question into a set of triples.

6.3 Simple Questions over KB

Simple questions can be answered by one fact from KB. The available models for solving this problem aim at extracting the single fact which answers the question. Answer of the question is stored as the tail entity in the target fact. The available models try to find the fact by detecting the head entity and relation from the question sentence and forming a SPARQL query. We will discuss the architecture of models proposed for simple QA over KB in the following.

Mohammed et al. (2018) show that deep learning structures used in simple QA systems are unnecessarily complicated. They divided the simple QA task into four subtasks, namely, entity detection, entity linking, relation prediction, and evidence integration. Each of these subtasks is explained below.

- **Entity detection:** The goal of this subtask is determining the entity being queried. In the output, each of the question words is labeled as ENTITY or NOENTITY. Two different approaches are proposed for entity detection including RNNs and conditional random fields (CRFs). The question word embeddings are passed to an RNN followed by a linear function. Each question word is mapped to a label in the last layer of this model. In the case of using bi-GRU or BiLSTM, two output hidden states in the forward and backward directions are concatenated and passed to the linear function. CRFs are also used for entity detection due to their great performance in sequence labeling tasks. The CRF model used in (Finkel et al., 2005) which incorporates features like word positions, POS tags, and character n-grams is utilized.
- **Entity linking:** The candidate entities are extracted in the previous step and are linked to a real node in the knowledge graph. Nodes are represented by a machine identifier in Freebase. A fuzzy string matching approach is utilized for linking the entities. An inverted index over n-grams of entity names in Freebase is generated. Given the candidate entities, their corresponding n-grams are created and searched to find matching. The matching process is started from higher

n-grams, and in the case of observing an exact match, the lower n-grams are not processed. A list of machine identifiers of retrieved relevant candidate entities is extracted and ranked by the Levenshtein distance.

- **Relation prediction:** The goal of this step is predicting the relation given the whole question sentence. Relation prediction is performed by using three models based on RNNs, CNNs, and logistic regression.
  RNNs: The input question is passed to BiLSTM or bi-GRU model, and the output hidden state corresponding to the last token is used for classification.
  CNNs: Feature maps with size of two, three, and four are applied to word embeddings of the input question. Max-pooling is applied on the output of feature maps, and a fully connected layer followed by softMax is used for classification.
  Logistic regression: Logistic regression is used for classifying the input question with two different sets of features. In the first setting, TF-IDF weights of unigrams and bigrams are used as features. In the second setting, the average of question word embeddings is concatenated to a one-hot vector with 300 dimensions. Each dimension represents one of the most frequent words from the relation names.
- **Evidence integration:** The goal of this step is to retrieve one relevant fact from the knowledge graph. A list of relevant entities and relations is extracted separately in the previous steps. Different combinations of entity and relation can be built by using the extracted entities and relations. They can be scored by multiplying the relevance score of entity and relation. Some of the entity relations do not exist in the knowledge graph and are pruned. Two different scoring ties are applied for scoring the remaining identity relation sets including (1) favoring popular entities which have more incoming edges in the knowledge graph and (2) favoring entities which have mapping to Wikipedia.

GENQA (Yin et al., 2016a) is an end-to-end answer generator (GENQA) based on a knowledge base to answer simple questions, which need a single fact from knowledge graph. The structure of the model is based on sequence2sequence approach. Since this model uses an end-to-end approach, it does not need linguistic knowledge. The output of the model contains two types of words: knowledge base words which are extracted from the knowledge base and common words which are used to complete the answer's sentence. As the architecture of GENQA is shown in Fig. 6.4, the main three components of GENQA are Interpreter, Enquirer, and Answerer. The system also benefits from an external knowledge base. Interpreter encodes the given question Q into a fixed-length vector representation H_Q and stores it in a short-term memory. Given the encoded question representation H_Q and external knowledge base, Enquirer extracts the relevant facts from the knowledge base and represents them by r_Q. The encoded question representation H_Q and the relevant facts r_Q are passed to Generator for generating the answer. Each of these components is described below:

- **Interpreter:** Input question $Q = (x_1, \cdots, x_{T_Q})$ is converted to a sequence of vector embeddings. Contextual representation of each question term is generated using bi-GRU which processes the input sentence in two forward and backward

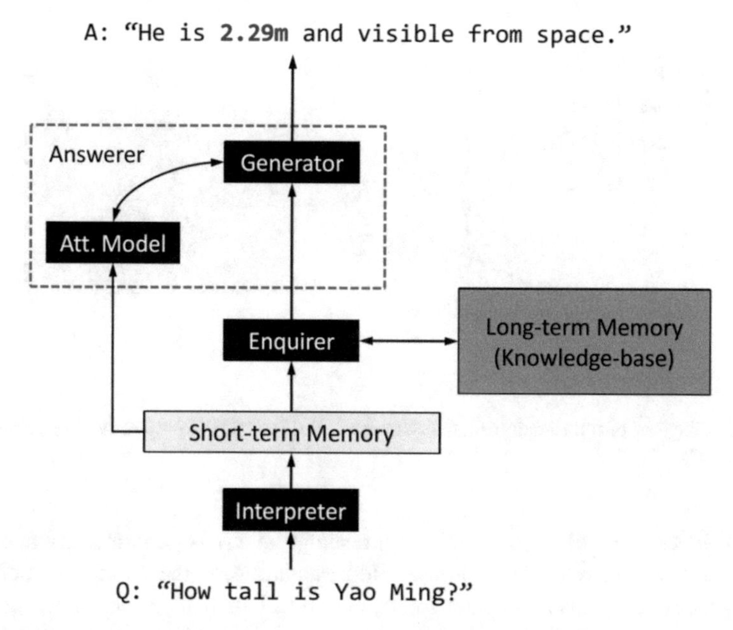

Fig. 6.4 The architecture of the GENQA model (Yin et al., 2016a) ("©1963–2022 ACL, reprinted with permission")

directions. Final representation of each question term ($\tilde{h}_i$) is generated by concatenating word representation (x_t), one-hot encoded representation (x_t), and corresponding hidden state generated by bi-GRU (h_i) as follows:

$$\begin{aligned} H_Q &= (\tilde{h}_1, \cdots, \tilde{h}_{T_Q}) \\ \tilde{h}_t &= [h_t; x_t; x_t] \end{aligned} \tag{6.5}$$

- **Enquirer:** The relevant candidate facts to the input question are extracted by term-level matching and represented as $\mathcal{T}_Q = \{\tau_k\}_{k=1}^{K_Q}$. The matching score between each candidate fact and input question is calculated using CNN-based matching or bilinear model. The diagram of the Enquirer model is shown in Fig. 6.5. The matching scores are stored in r_Q vector which is calculated as follows:

$$r_{Q_k} = \frac{e^{S(Q,\tau_k)}}{\sum_{k'=1}^{K_Q} e^{S(Q,\tau_{k'})}} \tag{6.6}$$

where r_{Q_k} is the k-th element of vector r_{Q_k} and $S(Q, \tau_k)$ is the matching score of question Q and candidate fact τ_k which could be measured by a bilinear or CNN-based matching model.

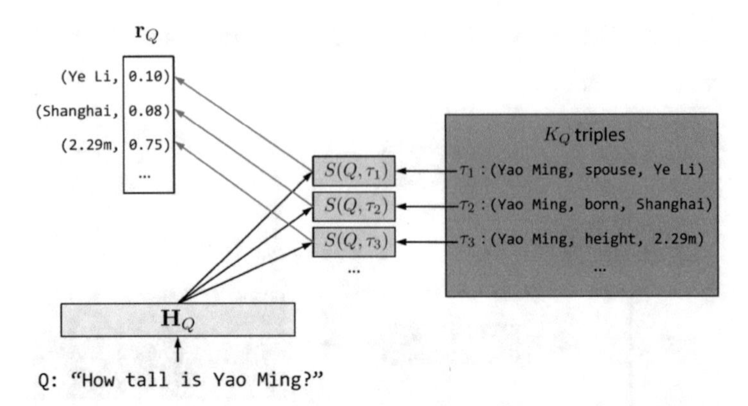

Fig. 6.5 Diagram of the Enquirer model (Yin et al., 2016a) ("©1963–2022 ACL, reprinted with permission")

- **Bilinear model:** The question representation $\overline{x}_Q$ is generated by averaging the question word embeddings. Representation of the candidate fact u_τ is generated by averaging the representation of its subject and predicate. The matching score of given question and candidate fact is calculated as follows:

$$\bar{S}(Q, \tau) = \overline{x}_Q^\top M u_\tau \tag{6.7}$$

  where M is the matching matrix.
- **CNN-based matching model:** A CNN composed of a convolution layer followed by a max-pooling layer is used for generating the representation of the input question. Representation of the candidate fact is generated similar to bilinear model u_τ. Representation of the question sentence is concatenated to representation of the candidate fact and passed to a MLP for measuring the matching score of them as follows:

$$\hat{S}(Q, \tau) = f_{\mathrm{MLP}}([\hat{h}_Q; u_\tau]) \tag{6.8}$$

- **Answerer:** An RNN is used for generating the answer sentence representation given the information extracted from question (H_Q) and the knowledge retrieved from knowledge base (r_Q). The diagram of the Answerer model is shown in Fig. 6.6. The probability of generating the answer sentence $Y = (y_1, y_2, \ldots, y_{T_Y})$ is calculated as follows:

$$p(y_1, \cdots, y_{T_Y}|H_Q, r_Q; \theta) = p(y_1|H_Q, r_Q; \theta) \prod_{t=2}^{T_Y} p(y_t|y_{t-1}, s_t, H_Q, r_Q; \theta) \tag{6.9}$$

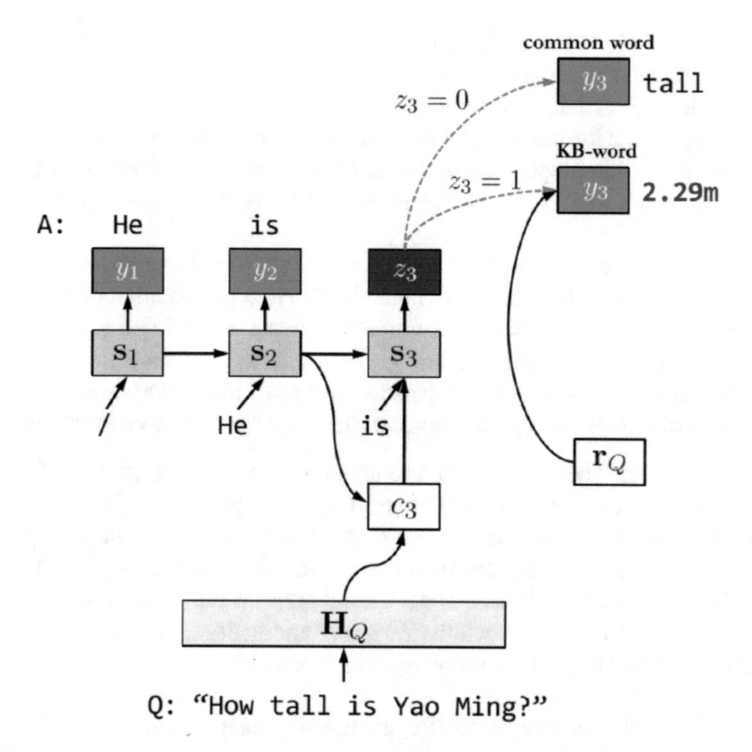

Fig. 6.6 Diagram of the Answerer model (Yin et al., 2016a) ("©1963–2022 ACL, reprinted with permission")

where s_t is the t-th hidden state and the probability of generating the t-th word is as follows:

$$p(y_t|y_{t-1}, s_t, H_Q, r_Q; \theta) = p(z_t = 0|s_t; \theta)p(y_t|y_{t-1}, s_t, H_Q, z_t = 0; \theta) + p(z_t = 1|s_t; \theta)p(y_t|r_Q, z_t = 1; \theta) \tag{6.10}$$

where the latent variable $z_t = 0$ represents word y_t is originated from common vocabulary and $z_t = 1$ represents word y_t is originated from the knowledge base. The knowledge base vocabulary is all of the objects from the candidate facts. To be more specific, each word from the answer sentence is generated by using both language and knowledge base parts with weight $p(z_t|s_t; \theta)$. This weight is generated by utilizing a logistic regression model given the hidden state s_t. When $z_t = 0$, the probability is generated by using an RNN decoder (Bahdanau et al., 2014). Hidden state at time step t is generated as $s_t = f_s(y_{t-1}, s_{t-1}, c_t)$ and $p(y_t|y_{t-1}, s_t, H_Q, z_t = 0; \theta) = f_y(y_{t-1}, s_t, c_t)$. The context vector c_t is the weighted sum of the hidden states stored in H_Q. When $z_t = 1$, $p(y_t = k|r_Q, z_t = 1; \theta) = r_{Q_k}$.

Sorokin and Gurevych (2017) proposed an end-to-end model which constructs semantic graphs according to a given question. Semantic graph is constructed by using a question variable, constrained, relation types, and entities. The question variable represents the answer of the question and must be replaced with an entity from the knowledge graph. The semantic graph is created according to the given knowledge graph and consequently is limited to relation types defined in that knowledge graph.

Encoded representations of the question and the candidate semantic graphs are constructed using a CNN-based architecture. Then, representation of the question is compared with encoded representation of candidate semantic graphs using cosine similarity. The most relevant semantic graph is selected and converted to SPARQL query for retrieving the answer of question from Wikidata RDF dump. A detailed description of step-by-step performance of this model is explained in the following.

- **Entity linking:** The input question sentence is tokenized, and its token fragments are extracted by a set of pre-defined regular expressions. Each fragment is converted to a set of possible n-grams. Wikidata entities include a set of alternative labels, and the obtained n-grams are searched among the alternative labels to find relevant entities. The retrieved relevant entities are ranked using the Levenshtein distance between the fragment and entity label and using the item identifier according to the following score function:

$$\begin{aligned}\text{rank } =& a \text{ levenshtein(fragment, entity\_main\_label)} \\ &+ b \log \text{entity\_serial\_id} \\ &+ c \max(1 - \frac{\text{len(entity\_label)}}{\text{len(fragment)}}, 0)\end{aligned} \tag{6.11}$$

  where a, b, and c are coefficients. In the third part of the above formulation, longer fragment matches are given a higher score.
- **Iterative representation generation:** Having the list of candidate entities from the previous step, they are used for generating representation graphs iteratively. The first graph is empty and initialized by a question variable. In each step of the iterative procedure, one of the actions for graph generation is conducted, and the new graph is added to the list of candidate graphs. The possible actions include ADD_RELATION, ADD_TEMPORAL_CONSTRAINT, and ADD_NUMBER_CONSTRAINT. Figure 6.7 shows the procedure of constructing a semantic graph. For applying each action, a set of pre-defined conditions must be checked.
- **Neural vector encodings:** The vector encoding of the question sentence and the candidate semantic graphs are obtained using a CNN-based neural architecture. Figure 6.8 represents the model used for encoding the question sentence. Named entities of the question sentence are replaced by a special token $\langle e \rangle$; the beginning and end of the sentence are marked with $\langle S \rangle$ and $\langle E \rangle$ tokens, respectively. The trigram embeddings of question words are passed to the convolution layer fol-

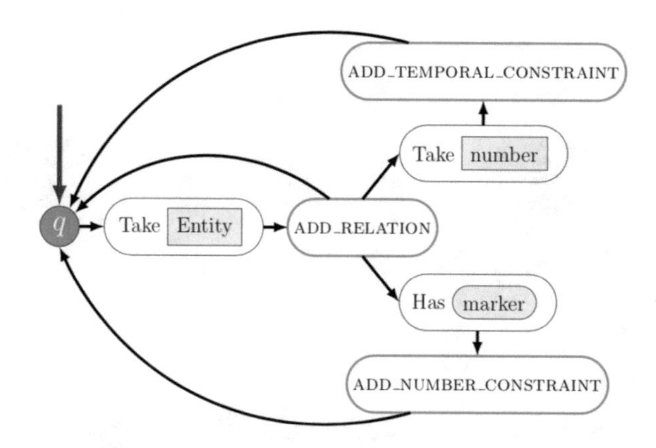

Fig. 6.7 Process of constructing a semantic graph (Sorokin & Gurevych, 2017) ("©2017 Springer International Publishing AG, reprinted with permission")

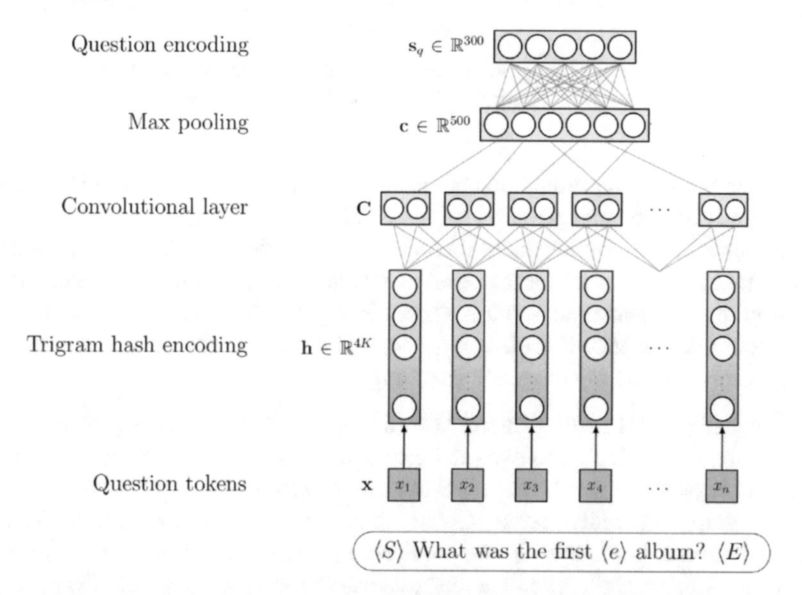

Fig. 6.8 The architecture of question encoder model (Sorokin & Gurevych, 2017) ("©2017 Springer International Publishing AG, reprinted with permission")

lowed by a max-pooling layer. Each word is represented with a binary vector with dimension V, where V is the count of distinct possible trigrams. For extracting trigrams of a given word, the token # is added to the beginning and end of the word. The word "what" contains set of trigrams $t = \{\#wh, wha, hat, ha\#\}$. Output of the pooling layer is passed to a fully connected layer for generating the encoding of the question.

The same neural architecture with shared weights is used for encoding the semantic graph. Each relation within the semantic graph is encoded separately,

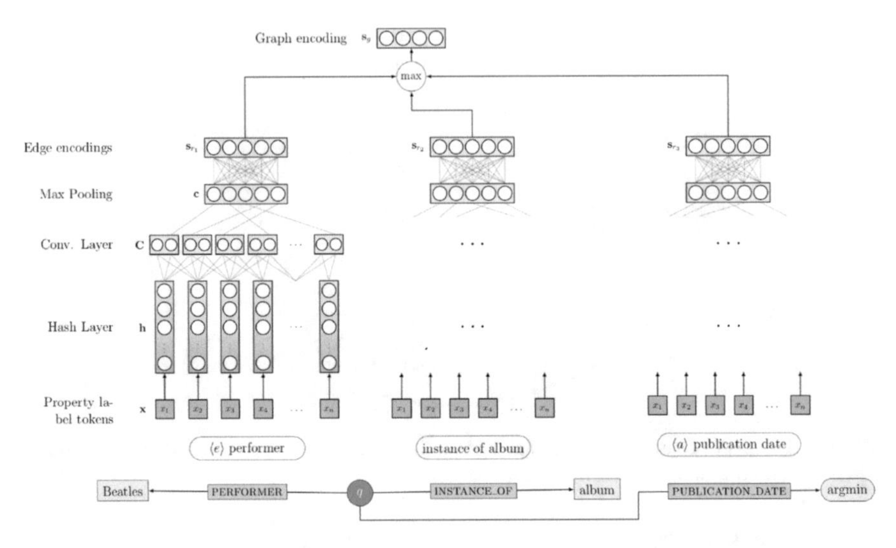

Fig. 6.9 An example of generating the encoding of a semantic graph (Sorokin & Gurevych, 2017) ("©2017 Springer International Publishing AG, reprinted with permission")

and the semantic representation of the given graph is constructed by max-pooling over relation encodings. A token $\langle e \rangle$ is added to the beginning or end of the relation label, depending on the direction of the relation. The label "point in time" and token $\langle a \rangle$ are used for temporal constraints. An example of constructing the encoding of a semantic graph is shown in Fig. 6.9. The output of this model is a set of relation encodings $\{s_{r_1}, s_{r_2}, \cdots, s_{r_m}\}$ which are passed to a max-pooling layer for constructing the graph encoding s_{r_q}.

Knowledge Embedding-based QA (KEQA) framework (Huang et al., 2019) embeds the given graph into two predicate and entity embedding spaces. Each fact from the knowledge graph (h, ℓ, t) would be represented as triples (e_h, p_ℓ, e_t) in the new graph embedding space. Given the question sentence, KEQA predicts the representation of head entity $\hat{e}_h$ and predicate $\hat{p}_\ell$. Then, the tail entity $\hat{e}_t$ is estimated by function $\hat{e}_t = f(\hat{e}_h, \hat{p}_\ell)$. The closest fact to the estimated fact $(\hat{e}_h, \hat{p}_\ell, \hat{e}_t)$ is retrieved from knowledge graph by using a joint distance metric. The architecture of the KEQA model includes several steps that are described in the following.

- **Knowledge graph embedding:** Knowledge graph embedding methods create an embedding for each node and vertex of the graph by preserving the original connections and relations in the graph. The knowledge graph embedding models use a function for predicting the tail entity given the head entity and predicate $e_t \approx f(e_h, p_\ell)$. Vector embeddings of entities and predicates are initialized randomly or with word embeddings and are learned by minimizing the difference between the actual embedding of head entity and predicted value. TransE (Bordes et al., 2013) and TransR (Lin et al., 2015) models predict the head entity embedding by using functions $e_t \approx e_h + p_\ell$ and $e_t M_\ell \approx e_h M_\ell + p_\ell$, respectively.

- **Predicate representation learning:** This component is used for obtaining the predicate embedding from the question. Word embeddings of question words are passed to two forward and backward LSTMs. Hidden state representation of each token is created by concatenating the corresponding forward and backward hidden state representations $h_j = [\overrightarrow{h}_j; \overleftarrow{h}_j]$. The attention weight is computed for each hidden state h_j as follows:

$$\alpha_j = \frac{\exp(q_j)}{\sum_{i=1}^{L} \exp(q_i)}$$
$$q_j = \tanh(w^{\top}[x_j; h_j] + b_q) \tag{6.12}$$

Using the attention weight α_j, the new representation is created for each token $s_j = [x_j; \alpha_j h_j]$ and passed to a fully connected layer for predicting representation of each token ($r_j \in \mathbb{R}^{d\times 1}$). Predicted encoding of predicate is generated by averaging r_j vectors as follows:

$$\hat{p}_\ell = \frac{1}{L}\sum_{j=1}^{L} r_j^{\top} \tag{6.13}$$

- **Head entity learning model:** The same architecture proposed for learning predicate embedding is used for inferring the head entity encoding $\hat{e}_h$ from input question. Comparing the predicted head entity encoding with all of the entity embeddings in E is not optimal as the knowledge graph includes a large number of head entities. To reduce the search space, a head entity detection model is employed for selecting a subset of candidate entities.
- **Head entity detection (HED) model:** HED model classifies each token of input sentence as entity or non-entity. Word embeddings of question terms are passed to a BiLSTM to produce the corresponding hidden states h_j. The hidden states are passed to a fully connected layer followed by a softMax for predicting the probability of each category by producing vector $v_j \in \mathbb{R}^{2\times 1}$. $\text{HED}_{\text{entity}}$ is extracted in this layer. All of the entities which are identical to $\text{HED}_{\text{entity}}$ or include $\text{HED}_{\text{entity}}$ are selected from the knowledge graph.
- **Joint distance metric:** Candidate facts are selected from the knowledge graph by comparing their name with $\text{HED}_{\text{entity}}$. A candidate fact which has the minimum distance with predicted fact $(\hat{p}_\ell, \hat{e}_h, \hat{e}_t = f(\hat{e}_h, \hat{p}_\ell))$, according to the following distance function, is selected as answer of the question:

$$\underset{(h,\ell,t)\in C}{\text{minimize}} \|p_\ell - \hat{p}_\ell\|_2 + \beta_1\|e_h - \hat{e}_h\|_2 + \beta_2\|f(e_h, p_\ell) - \hat{e}_t\|_2$$
$$- \beta_3 \,\text{sim}[n(h), \text{HED}_{\text{entity}}] - \beta_4 \,\text{sim}[n(\ell), \text{HED}_{\text{non}}] \tag{6.14}$$

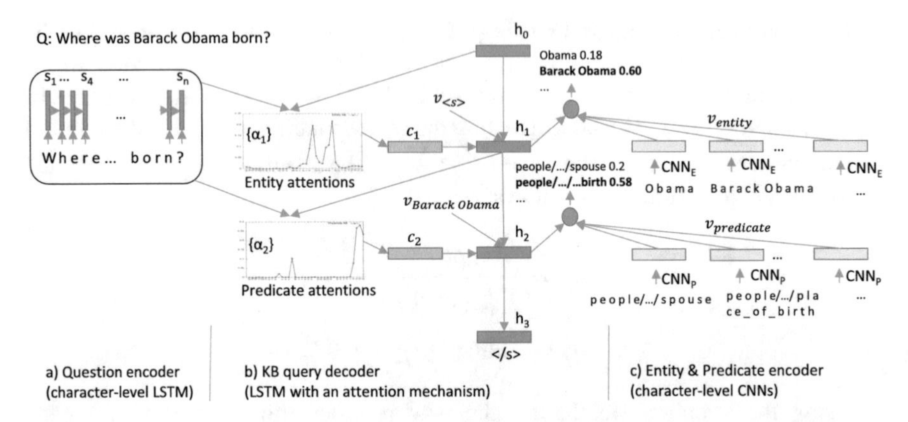

Fig. 6.10 The architecture of the model proposed by He and Golub (2016) ("©1963–2022 ACL, reprinted with permission"). (**a**) Question encoder (character-level LSTM). (**b**) KB query decoder (LSTM with an attention mechanism). (**c**) Entity and predicate encoder (character-level CNNs)

where function n gives the name of input entity or predicate, function sim calculates the similarity of two input strings, distance of vectors is calculated by using ℓ^2 norm, and β_1, β_2, β_3, and β_4 are the weights of each term.

He and Golub (2016) proposed a model that converts each question into a knowledge base query which includes topic entity and predicate. The pair of topic entity and predicate is selected by using the function f given the question (q), set of candidate predicates ($\{e\}$), and entities ($\{p\}$). Function f estimates the probability of generating topic entity and predicate given the question sentence. As the architecture of this model is shown in Fig. 6.10, it includes three components which are described in the following.

- **Encoding the question:** The question sentence is encoded by feeding the character-level one-hot encoding vectors to a two-layer LSTM. The character-level representations are represented by $x_1, \cdots, x_n$, and output of LSTM is denoted by $s_1, \cdots, s_n$ vectors.
- **Encoding entities and predicates in the knowledge base:** Two distinct convolutional neural networks are used for encoding the entities and predicates in the knowledge base. The neural model consists of two successive convolution layers followed by a feed-forward layer as follows:

$$f(x_1, \ldots, x_n) = \tanh(W_3 \times \max(\tanh(W_2) \operatorname{conv}(\tanh(W_1 \times \operatorname{conv}(x_1, \ldots, x_n))))) \tag{6.15}$$

where max indicates the max-pooling operation and conv is the temporal convolution layer.

- **Decoding the knowledge base query:** An LSTM-based neural model is utilized for decoding the entities/predicates. The output of LSTM at each time step is compared with embedding of other entities/predicates to find the most similar one by using a semantic relevance function. An attention-based LSTM is used for decoding the input and generating the attentional representation of the question sentence as follows:

$$c_i = \sum_{j=1}^{n} \alpha_{ij} s_j$$
$$\alpha_{ij} = \frac{\exp(e_{ij})}{\sum_{k=1}^{T_x} \exp(e_{ik})} \tag{6.16}$$
$$e_{ij} = v_a^\top \tanh(W_a h_{i-1} + U_a s_j)$$

where c_i is the context vector which indicates the attentional representation of the question at time step i, h_i is the hidden state of LSTM at time step i, s_j is the embedding of the j-th character encoding with size r, and W_a, U_a, and v_a are the model parameters.

The hidden states, produced by decoder, are compared with entity and predicate encodings to measure the semantic similarity of the question with predicate and entities in the knowledge base. The probability of producing entity e_j and predicate p_k is calculated as follows:

$$P(e_j) = \frac{\exp(\lambda \cos(h_1, e_j))}{\sum_{i=1}^{n} \exp(\lambda \cos(h_1, e_i))}$$
$$P(p_k) = \frac{\exp(\lambda \cos(h_2, p_k))}{\sum_{i=1}^{m} \exp(\lambda \cos(h_2, p_i))} \tag{6.17}$$

where h_i is i-th hidden state of LSTM, $e_1, \cdots, e_n$ are the embedding of entities, $p_1, \cdots, p_m$ are the embedding of predicates, and λ is a constant. The probability of each pair of entity and predicate given the question sentence is calculated by maximizing the following function:

$$(e^*, p^*) = \text{argmax}_{e_i, p_j}(P(e_i) * P(p_j)) \tag{6.18}$$

where e^*andp^* are the corresponding entities and predicates in the knowledge base to question q and are selected from a set of candidate entities $\{e\}$ and predicates $\{p\}$. The English alias of candidate entities is a substring of the input question. The candidate entities whose alias is a substring of another entity are eliminated. All the predicates belonging to candidate entities are selected to form the set of candidate predicates.

Yin et al. (2016b) proposed a model where two different entity linking approaches, namely, passive entity linker and active entity linker, are used. In passive entity linker, the question words are used for searching the candidate entities, and mention is detected using the candidate entities. In active entity linker,

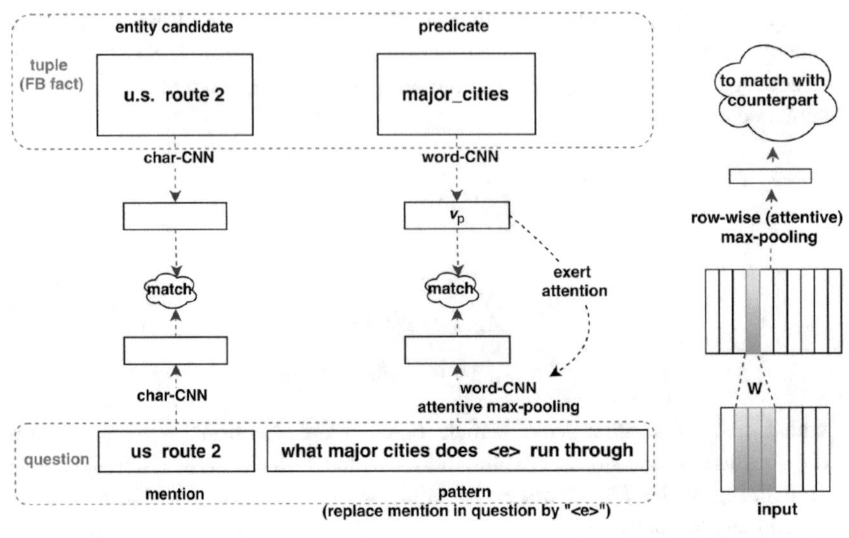

(a) The whole system. Question: what major cities does us route 2 run through; Tuple: ("u.s. route 2", "major_cities")

(b) Convolution for representation learning

Fig. 6.11 The architecture of the model proposed by Yin et al. (2016b) ("©1963–2022 ACL, reprinted with permission"). (**a**) The whole system. Question what major cities does us route 2 run through; Tuple: ("u.s route 2", "major_cities"). (**b**) Convolution for representation learning

mention is detected in the question sentence at first, and the candidate entities are selected using the span of mention. Different neural models are proposed for mention detection.

The architecture of this model is shown in Fig. 6.11. Inputs of the neural model are (subject, predicate) from knowledge base and (mention, pattern) from question sentence. The pattern is generated by replacing the mention with $< e >$ token in the question sentence. After the entity linking step, fact selection is applied to candidate facts from the knowledge base. The candidate entities from the knowledge base are compared with mention string by using a CNN on character-level (char-CNN) and predicates are compared with pattern via a CNN with attentive max-pooling on word-level (word-AMPCNN). The word-CNN model is also used for creating the representation of the predicate from the knowledge base fact. Both the word-CNN and char-CNN models share the same convolution weights and have the same architecture which is described in the following.

- **Input layer:** Input of this model is a matrix with $d \times s$ dimension where s is the maximum sequence length of input and d is the dimensionality of tokens' embedding.
- **Convolution layer:** Convolution is applied on the sequence of n consecutive tokens. Input of this layer is a sequence of s tokens $v_1, v_2, \cdots, v_s$, and vector c_i is generated by concatenating the n consecutive tokens $v_{i-n+1}, \cdots, v_i$. The

output vector p_i of convolution is generated as follows:

$$p_i = \tanh(W \cdot c_i + b) \tag{6.19}$$

where $W \in \mathbb{R}^{d \times nd}$ is the convolution weight and $b \in \mathbb{R}^d$ is the bias.

- **Max-pooling:** Max-pooling is applied on output vectors p_i to produce the representation of j-th token of sequence s as follows:

$$s_j = \max(p_{j1}, p_{j2}, \cdots) \tag{6.20}$$

An example of AMP is shown and compared with traditional max-pooling (TMP) in Fig. 6.12. In AMP-CNN, the feature map of pattern (F_{pattern}) represents the local n-grams in each column. The feature map is compared with the representation of predicate by using cosine similarity as follows:

$$s_i = \cos(v_p, F_{\text{pattern}}[:, i]) \tag{6.21}$$

The re-weighted feature map (F_{decay}) is generated as follows:

$$F_{\text{decay}}[:, i] = F_{\text{pattern}}[:, i] * \overline{s}_i \tag{6.22}$$

where $\overline{s}$ is obtained by changing negative values of s to zero and normalizing the positive values by dividing them to the largest value of s. Finally, the values from the first feature map F_{pattern} with the same coordinate of max values from matrix F_{decay} are selected.

Based on our review on the models proposed for simple QA over KB, we observe a high degree of variation in the methods utilized for solving this task. In the model proposed by Mohammed et al. (2018), the entity and the relation are predicted using the question sentence. The different permutations of entities and relations are used for extracting the fact that answers the question from KB. GENQA is an end-to-end model that encodes the input query into a vector representation and extracts the relevant facts to the encoded representation from KG. GENQA selects the most relevant fact by using a CNN-based or a bilinear model for measuring the matching score of question encoding and the relevant facts. The answer sentence is generated using the fact and the question sentence. Sorokin and Gurevych (2017) developed CNN-based models to construct semantic graphs from the input question and represent the input question. The question's representation is compared with encoded representation of the candidate semantic graphs to find the semantic graph which answers the question. The KB is queried using the semantic graph to retrieve the answer. Despite the other models which only used the KB for querying, the KEQA model represents the entities and the relations of the KB in a new embedding space. In the new space, given the head entity and the relation, representation of the tail entity can be obtained. KEQA employs two networks for detecting the head entity and predicate representations from the question sentence. Then the tail entity is obtained as a function of head entity and relation. He and Golub (2016) used the

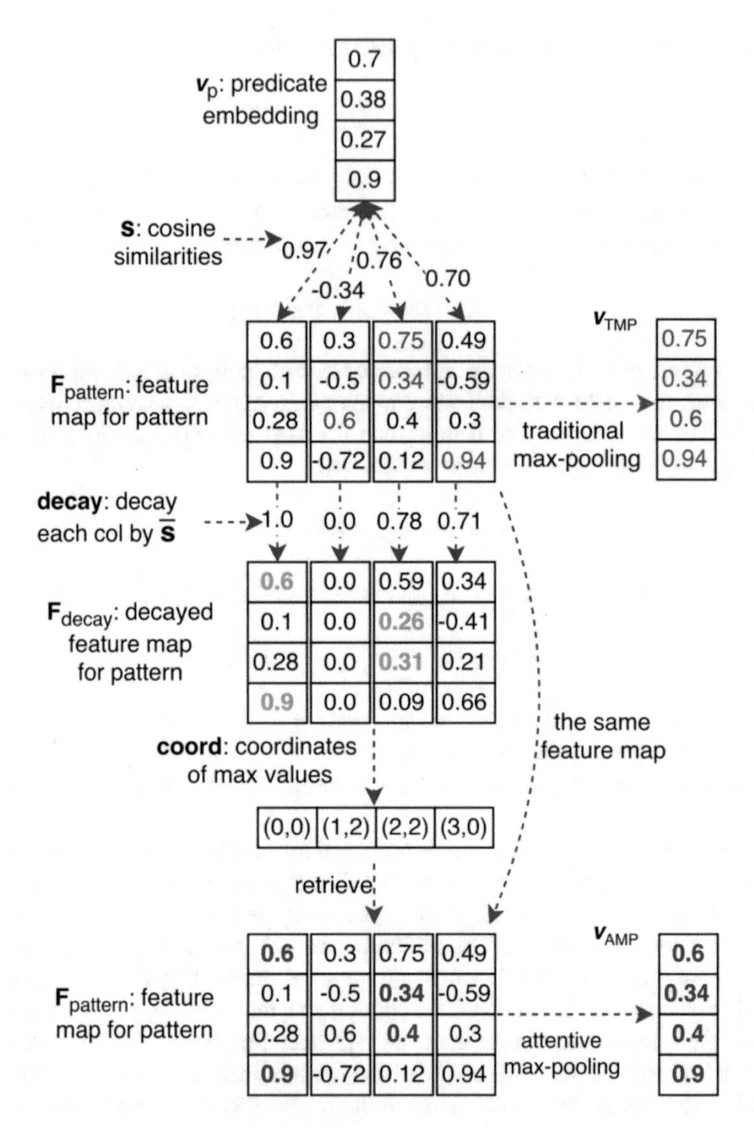

Fig. 6.12 Illustration of traditional max-pooling (TMP) vs attentive max-pooling (AMP) Yin et al. (2016b) ("©1963–2022 ACL, reprinted with permission")

character-level encoding of the question sentence. They approached the problem as finding the topic entity and predicate from question sentence. They produced the embedding of KB entities and predicates. These embeddings and the embedding of question were then used to find the set of most probable candidate predicates and entities. The model proposed by Yin et al. (2016b) detects the candidate entities and mention from question sentence and then performs entity linking and fact selection.

The model used character-level and word-level CNN for comparing the entities from KB with candidate entities and predicates from KB with mention.

6.4 Complex Questions over KB

Answering the complex questions requires multiple facts from KB. As a result, the models proposed for simple QA cannot be used for answering the complex questions. We will discuss the architecture of some models proposed for complex QA in the following.

Hao et al. (2019) proposed a model using directed acyclic graph (DAG) embeddings for modeling the representation of question and knowledge base facts. As its architecture is shown in Fig. 6.13, this model includes four main components. The question sentence is encoded, and the candidate DAGs are extracted from the knowledge base. The candidate DAGs are encoded and compared with the encoded representation of the question to find the most relevant DAG which answers the given question. Each component of this model is described in the following.

- **Question encoder:** The contextual representation of the input question is modeled using BiLSTM. Last hidden states of both forward and backward LSTMs are concatenated to form the representation of the question.
- **DAG generator:** Generation of candidate DAGs is shown in Fig. 6.14. The question entities are extracted from the question sentence to form set $\mathcal{E}_q$; the entities that are linked to the question entities are also selected to form the

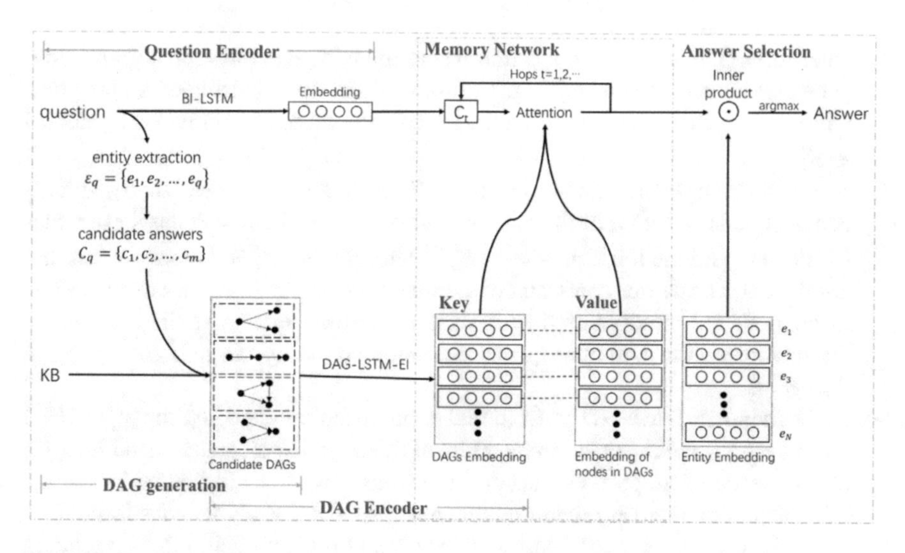

Fig. 6.13 The architecture of the model proposed by Hao et al. (2019) ("©1963–2022 ACL, reprinted with permission")

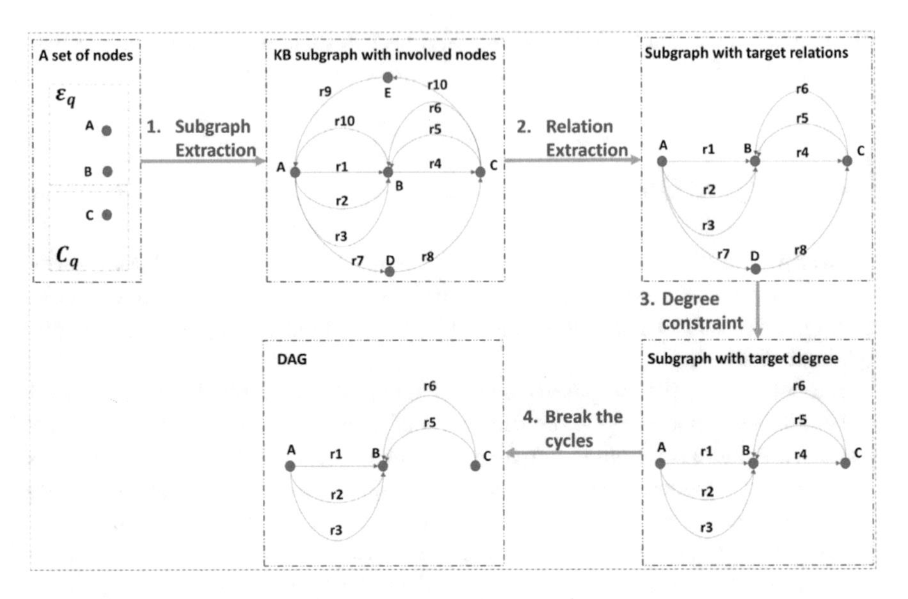

Fig. 6.14 An example of DAG generating process Hao et al. (2019) ("©1963–2022 ACL, reprinted with permission")

candidate answer set C_q. Similarity of each question q and relation r is calculated using the following equation:

$$s_{qr} = \max_{j=1,\ldots,l-N+1} (\phi(w_j w_{j+1} \ldots w_{j+N-1}) \cdot \varphi(r)) \tag{6.23}$$

where $q = \{w_1, w_2, \cdots, w_l\}$, ϕ returns the embedding of question n-gram and φ generates the relation embedding. The above similarity is calculated for the given question and each relation within the graph and first top k relations are selected as $R_q^{\text{top k}}$.

In the example provided in Fig. 6.14, $\mathcal{E}_q$ includes two nodes $\{A, B\}$, and C_q contains node $\{C\}$. Part of the main knowledge base which includes these nodes is selected, and the top k relations ($R_q^{\text{top k}}$) are preserved, and other vertices are eliminated. The degree constraint with limit of 3 is applied, and as a result, r_7 and r_8 relations are removed as they have less question-relation similarity. Finally, the relations with minimum question-relation similarity are deleted to remove the cycles.

- **DAG encoder:** DAG-LSTM-EI model is proposed for encoding the input DAG into a vector $g \in \mathbb{R}^d$. As the architecture of DAG-LSTM-EI is shown in Fig. 6.15, the sequence of graph nodes according to their order in the first node and the relation among them are fed to this model. DAG-LSTM-EI includes forget (f), input (i), output gates (o), hidden state (h), and memory cell ($\hat{C}$). The relation

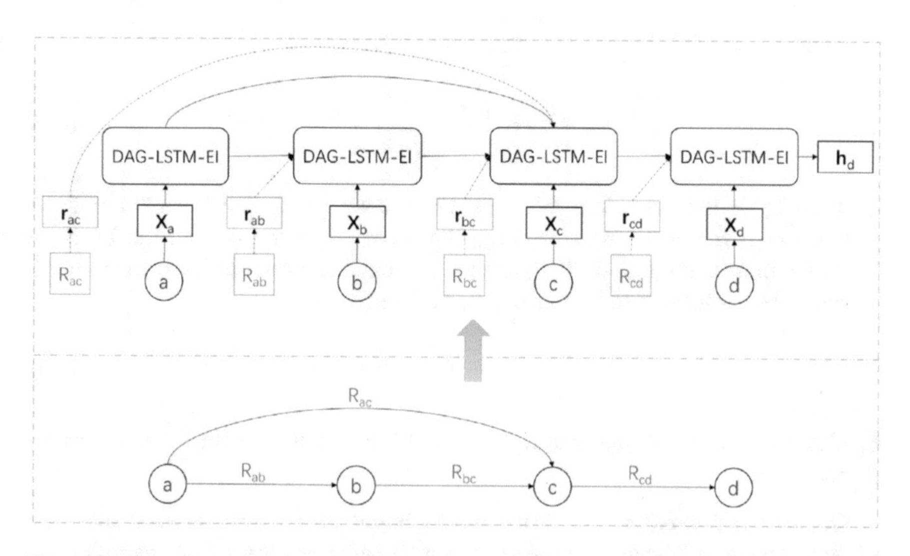

Fig. 6.15 The architecture of DAG-LSTM-EI model with an example of input DAG Hao et al. (2019) ("©1963–2022 ACL, reprinted with permission")

embeddings are fed as input of each time step, and formulation of this model is as follows:

$$
\begin{aligned}
i_j &= \sigma(W_i x_j + U_i(\sum_{j_k \in K_j} h_{j_k} + r_{j_k})) \\
f_j &= \sigma(W_f x_j + U_f(\sum_{j_k \in K_j} h_{j_k} + r_{j_k})) \\
\hat{C}_j &= i_j \times \tanh \sigma(W_u x_j + U_u(\sum_{j_k \in K_j} h_{j_k} + r_{j_k})) + \underset{j_k \in K_j}{\text{meanPool}}\{f_j \times (\hat{C}_{j_k} + r_{j_k})\} \\
o_j &= \sigma(W_0 x_j + U_0(\sum_{j_k \in K_j} h_{j_k} + r_{j_k})) \\
h_j &= o_j \times \tanh(S_j)
\end{aligned} \tag{6.24}
$$

where r_{j_k} is the relation connecting node j to node k, K_j represents the predecessors of node j, and $W \in \mathbb{R}^{d \times n}$ and $b \in \mathbb{R}^{d \times d}$ are the model parameters to be learned. The output of the last node is considered as the representation of input DAG.

- **Memory network:** A memory network which contains key-value pairs is used for storing the candidate DAG embedding (key) and nodes in the DAG (value). An attention layer is used for creating a representation of the question which is based on important question terms. The contextual representation of the question based on each DAG computed at time step t is calculated as follows:

$$c_t = W_t(c_{t-1} + W_p \sum_{(g,v)\in\mathcal{M}} (c_{t-1} \cdot g)v_i) \tag{6.25}$$

where g is the candidate DAG, v_i represents the embedding of a node in DAG, c_0 is the question embedding, W_p is the projection matrix, and W_t is the weight matrix. Finally, similarity of the contextual representation with entities is calculated as follows to find the answer of the question:

$$\text{Answer} = \arg\max_{a_i \in \mathcal{E}}(c_t \cdot v_{a_i}) \tag{6.26}$$

where v_{a_i} represents the embedding of candidate answer a_i and $\cdot$ represents the dot product.

QAmp (Vakulenko et al., 2019) works based on an unsupervised message-passing algorithm. The main feature of this approach is separating the reasoning and question interpretation processes. Besides that, edge direction is not considered in the QAmp model. QAmp includes two main steps, namely, question interpretation and answer inference. A set of entities and predicates required for answering the question with their confidence score are extracted in the question interpretation step. In the answer inference step, the confidence scores are propagated over the knowledge graph for finding the possible answers with their probability distribution. Each of these steps is described in the following.

- **Question interpretation:** The question model q is generated in two successive steps called parsing and matching. In the parsing step, the question sentence is parsed for extracting the references, including entities, predicates, and class motions, and recognizing the type of question. In the matching step, a list of candidate entities, predicates, and classes from the knowledge base with their rank is associated with references. Compound complex questions are answered by several questions which are connected to each other with their answers. To be more specific, the answer of one question is used as input for answering the other question. A chain of questions, which must be answered in sequence, are converted to hops from the knowledge base. The question model is defined as $q = \langle t_q, Seq_q \rangle$ where $t_q \in T$ represents the type of question Q (question types include {SELECT, ASK, COUNT}) and $\text{Seq}_q = (\langle E^i, P^i, C^i \rangle)_{i=1}^{h}$ indicates the set of h hops. Each hop is represented by entity E, property P, and class C.

  This question requires a set of two hops for answering:

  (1) finding the car types assembled in Broadmeadows Victoria, which have a hardtop style,
  (2) finding the company, which produces these car types.

and is modeled as follows:

$$
\begin{aligned}
q = \langle \text{SELECT}, (\\
&\langle E^1 = \{\text{"hardtop"}, \text{"Broadmeadows, Victoria"}\}, \\
&P^1 = \{\text{"assembles"}, \text{"style"}\}, \\
&C^1 = \{\text{ "cars" }\}\rangle, \\
&\langle E^2 = \emptyset, \\
&P^2 = \{\text{"company"}\}, \\
&C^2 = \emptyset\rangle)\rangle
\end{aligned}
\tag{6.27}
$$

In the parsing step, predicting the type of the question is a classification problem, and it is predicted by using a supervised classification model which assigns a type to each question. The supervised classification model is trained on a dataset of question and question type pairs. Another supervised sequence labeling model is used for reference extraction.

In the matching step, a set of relevant entities with their confidence score are retrieved from knowledge graph for each reference, and the question model is updated by function $I(q) = \left(t_q, SEQ_q\right)$. The property for the first hob is changed to $P^1 = \left\{P_1^1, P_2^1\right\} \in SEQ_q$ where $P_1^1 = \{$(dbo:assembly, 0.9), (dbp : assembly , 0.9)$\}$ and $P_2^1 = \{$(dbo: bodystyle , 0.5)$\}$.

- **Answer inference:** The confidence scores over the graph are aggregated by iteratively traversing the initial graph for concluding the answer. The answer set A^i, which contains a list of entities with their corresponding confidence score, is generated at the end of each hub and passed to the next hub plus the terms matched as input $\text{SEQ}_q(i+1) = \left\langle E^{i+1}, P^{i+1}, C^{i+1}\right\rangle$.

The last list of entities A^i is used for generating the answer through an aggregation function f_{t_q} dedicated to the question type t_q ($A_q = f_{t_q}\left(A^h\right)$). To sum up, the answer set at each hub is generated in two phases, namely, subgraph extraction and message passing phases.

1. **Subgraph extraction:** In this step, the relevant triples are extracted from the knowledge graph to create a subgraph. The entities and predicates of the query are matched with the knowledge graph entities and predicates, and the URIs of the matched items are selected to form a list of relevant entities and relations. The triples of the knowledge graph that contain at least one entity and one predicate from this list are selected.

   The subgraph formed in this step has n entities including all of the E^i entities and the entities adjacent to them via the P^i properties. The subgraph is shown with a three-dimensional matrix named $\mathbb{S}^{k \times n \times n}$. The matrix $\mathbb{S}$ contains a set of k-dimensional matrices with size $n \times n$ where n is the number of entities, k is the number of matched URIs, and each of the two-dimensional matrices corresponds to a URI type.

The n_{ij} cell of these two-dimensional matrices determines that an edge exists between the i-th and j-th entities with $n_{ij} = 1$. These matrices are symmetric as the edges are not directional, and as the entities are not connected to themselves, the diagonal values are zero.

2. **Message passing:** The main goal of this step is to propagate the confidence scores from the entities and predicates recognized at the interpolation step to the adjacent entities from the extracted subgraph. This algorithm is performed in three steps by (1) updating the properties, (2) updating the entities, and (3) aggregating the scores. The algorithm of message passing is summarized in Algorithm 4.

Algorithm 4 Message passing in KBQA (Vakulenko et al., 2019)

Input: adjacency matrices of the subgraph $\mathbb{S}^{k\times n\times n}$, entity $\mathbb{E}^{l\times n}$ and property reference activations $\mathbb{P}^{m\times k}$.
Output: answer activations vector $A \in \mathbb{R}^n$
$W^n, N_P^n, \mathbb{Y}_{\mathbb{E}}^{l\times n} = \emptyset$
for $P_j \in \mathbb{P}^{m\times k}, j \in \{1, \cdots m\}$ **do**
  $\mathbb{S}_j = \bigoplus_{i=1}^{k} P_j \otimes \mathbb{S}$ (Property update)
  $\mathbb{Y} = \mathbb{E} \oplus \otimes \mathbb{S}_j$ (entity update)
  $W = W + \bigoplus_{i=1}^{l} \mathbb{Y}_{ij}$ (sum of all activations)
  $N_{P_j} = \sum_{i=1}^{l} 1$ if $\mathbb{Y}_{ij} > 0$ else 0
  $\mathbb{Y}_{\mathbb{E}} = \mathbb{Y}_{\mathbb{E}} \oplus \mathbb{Y}$ (activation sums per entity)
end for
$W = 2 \cdot W/(l+m)$ (activation fraction)
$N_E = \sum_{i=1}^{l} 1$ if $\mathbb{Y}_{\mathbb{E}_{ij}} > 0$ else 0
return: $A = (W \oplus N_E \oplus N_P)/(l+m+1)$

The matches with a low confidence score are eliminated to keep a threshold for selecting the candidate answers. In addition, the answers can be filtered by selecting the entities of the answer set A^i that belong to one of the C^i classes.

SPARQL Query Generator (SQG) (Zafar et al., 2018) is a scalable approach which is capable of processing a noisy input. SQG is proposed to overcome some of the existing challenges in the query generation task. These challenges include large-scale KB, recognizing the question type, robust to noisy annotations, answering complex questions, and syntactic ambiguity. The architecture of SQG is shown in Fig. 6.16. We assume the entity and relation linking given the question sentence s are performed in the previous steps. The goal of SQG is to generate a set of candidate SPARQL queries for the input question and ranking them. Each question is modeled as a walk in knowledge graph which includes the entities E and relations R of the

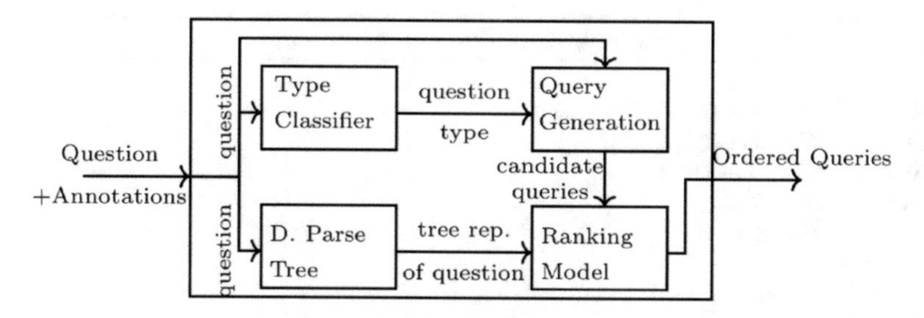

Fig. 6.16 The architecture of the SQG model (Zafar et al., 2018) ("©2018 Springer International Publishing AG, part of Springer Nature, reprinted with permission")

input question q as well as answer nodes. A walk and a valid walk in knowledge graph are defined as follows:

Definition 6.1 (Walk). A walk in a knowledge graph $K = (E, R, T)$ is a sequence of edges along the nodes they connect: $W = (e_0, r_0, e_1, r_1, e_2, ..., e_{k-1}r_{k-1}, e_k)$ with $(e_i, r_i, e_{i+1}) \in T$ for $0 \leq i \leq k-1$.

Definition 6.2 walk W is valid with respect to a set of entities E' and relations R', if and only if it contains all of them, i.e.:
$\forall e \in E' : e \in W$ and $\forall r \in R' : r \in W$

A node within W which does not belong to E' is named unbounded. Each component in the architecture of SQG is described in the following.

- **Query generation:** The subgraph of question is created given the knowledge graph, linked entities, and relations. By using the relevant subgraph, searching for valid walks is very time-consuming due to the limited size of the subgraph. The candidate valid walks are extracted from subgraph and converted to SPARQL query. Subgraph of a given question is created according to Algorithm 5. The subgraph includes all of the linked entities E which are expanded by linked relations R. Edges of the existing subgraph are extended by candidate relations except the corresponding relation of that edge. An example of a subgraph generated for question "What are some artists on the show whose opening theme is Send It On?" is shown in Fig. 6.17. As can be seen, this subgraph also includes the unbounded nodes in two-hop distance.

  Having the subgraph shown in Fig. 6.17, four valid walks based on Definition 6.2 are retrieved. The valid walks are represented in Fig. 6.18. When the extracted subgraph just has one valid walk, the only walk is converted to a SPARQL query.

  The query generator needs the type of the question for modifying the output SPARQL query accordingly. To this aim, questions are classified using an SVM and naïve Bayes model by their TF-IDF representation. Input questions are classified into three classes including list, count, and Boolean.
- **Query ranking:** A tree-LSTM model creates the representations of the candidate walks (Fig. 6.19) and the question (Fig. 6.20). Tree-LSTM takes the tree structure

Algorithm 5 Capturing the subgraph (Zafar et al., 2018)

Data: E', R', K
Result: G: Minimal covering subgraph
Initialize: G as an empty graph;
Add $\forall e \in E'$ to G as nodes;
for $e \in E', r \in R'$ **do**
  if $(e, r, ?) \in K$ **then**
    add $(e, r, ?)$ to G
  else if $(?, r, e) \in K$ **then**
    add $(?, r, e)$ to G
  end if
end for
for $(e_1, r, e_2) \in G$ **do**
  for $r' \in R', r' \neq r$ **do**
    if $(e_2, r', ?) \in K$ **then**
      add $(e_2, r', ?)$ to G
    else if $(?, r', e_2) \in K$ **then**
      add $(?, r', e_2)$ to G
    else if $(e_1, r', ?) \in K$ **then**
      add $(e_1, r', ?)$ to G
    else if $(?, r', e_1) \in K$ **then**
      add $(?, r', e_1)$ to G
    end if
  end for
end for

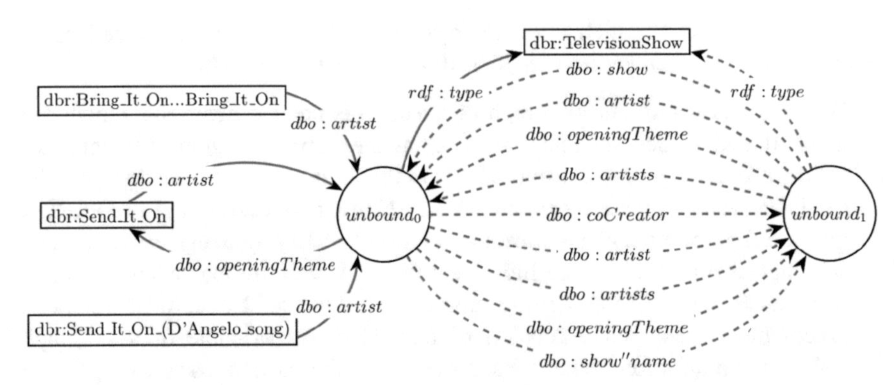

Fig. 6.17 Example of generated subgraph for the mentioned question. The solid lines represent one-hop distance, and dash lines represent two-hop distance Zafar et al. (2018) ("©2018 Springer International Publishing AG, part of Springer Nature, reprinted with permission")

as input. The dependency parse tree of the question (as shown in Fig. 6.20) is generated, and the entities within the parse tree are replaced by a placeholder. Tree representations of each candidate valid walk (is shown in Fig. 6.19) and question are fed to two distinct tree-LSTMs to create their latent representation. The latent representations of the question and candidate walk are compared by

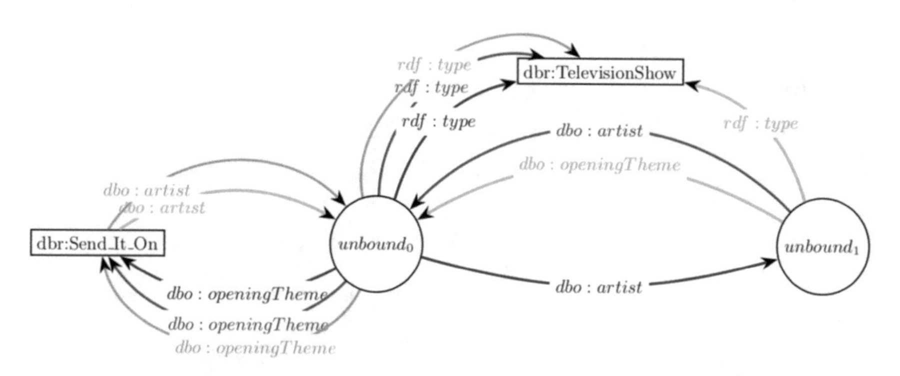

Fig. 6.18 Four candidate valid walks retrieved from the mentioned subgraph (Zafar et al., 2018) ("©2018 Springer International Publishing AG, part of Springer Nature, reprinted with permission")

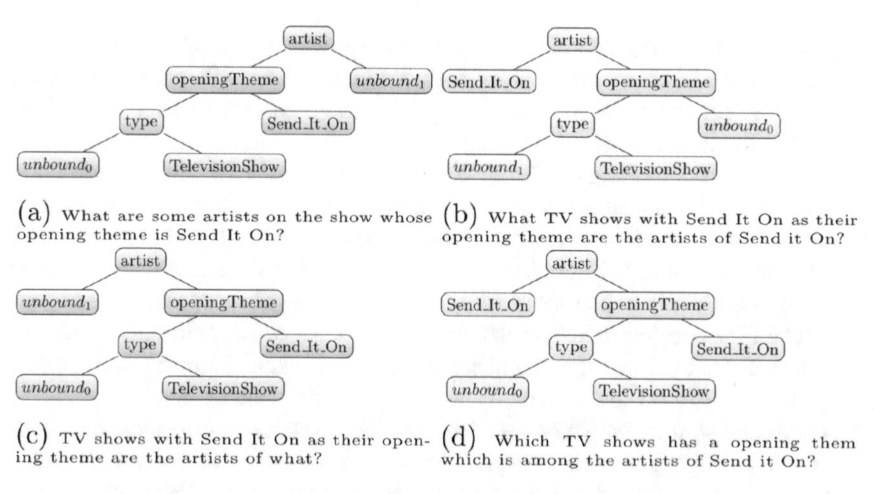

Fig. 6.19 Tree representation of candidate walks (Zafar et al., 2018) ("©2018 Springer International Publishing AG, part of Springer Nature, reprinted with permission")

using a similarity function. Finally, the candidate walks are ranked according to their similarity.

Zhu et al. (2020) proposed a model for KBQA based on tree-to-sequence learning. At the first step, constraint linking is performed for finding the possible entities which the question refers to. At the second step, candidate queries are generated using the constraints obtained in the previous step. At the third step, the candidate queries are encoded, and the encoded representations of queries are decoded in the fourth step to find the most relevant query. Finally, the most relevant query is used for querying the knowledge graph and finding the answer. Each of these steps is explained in the following.

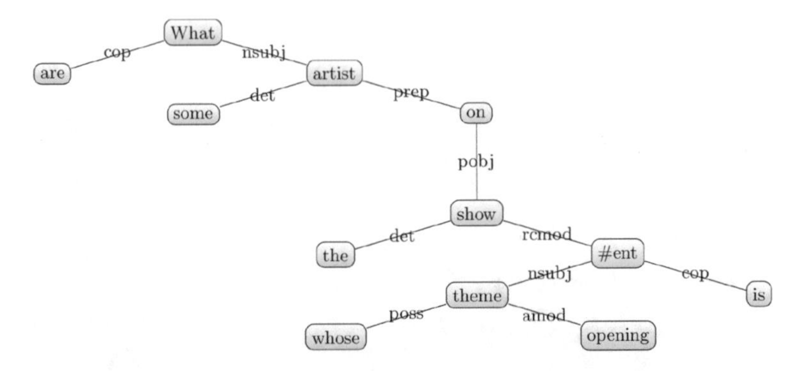

Fig. 6.20 Parse tree of input question (Zafar et al., 2018) ("©2018 Springer International Publishing AG, part of Springer Nature, reprinted with permission")

- **Constraint linking:** Part of the knowledge graph which is possibly mentioned by the question is selected for further steps. Entities, types, and number constraints are used for extracting part of the knowledge graph. For linking the entities within the question with knowledge graph entities, a novel entity linking tool named S-MART (Yang & Chang, 2015) is utilized. S-MART works by finding the mentions of the question and assigning a set of entities to them with a specific confidence score. A set of available distinct types from Freebase are selected and used for detecting the question types. The types that include a question word or phrase are selected as question type. By using regular expressions and pre-defined lexicon, the numbers that are required for constructing the query are selected. In addition, single numbers, comparative tokens followed by a number (e.g., before 1998), and superlative words (e.g., oldest) are extracted.
- **Candidate query construction:** The candidate queries are generated in four steps by using the constraints captured in the previous step. (1) The linked entities are leaf nodes, and other entities in one-hop distance of the linked entities are selected as intermediate nodes. The answer of the question is an intermediate node. (2) The intermediate nodes are linked to the graph. (3) An intermediate node is considered as an answer variable, and the remaining of the intermediate variables are considered as regular variables. In the case of selecting a CVT node as answer variable, the other node in one-hop distance of CVT is considered as an answer variable (CVT is a special node in Freebase used for combining a set of nodes to form a fact). (4) When an intermediate node includes number or type operations, the number and type attributes are queried and attached to the intermediate nodes before converting them to variables.
- **Tree-based encoder:** A candidate query generated for question "What character did Liam Neeson play in Star Wars?" is shown in Fig. 6.21 with red color. An example of a tree-based encoder for the illustrated candidate query is represented in Fig. 6.22. Nodes of the graph are constructed by entities, types, number operations, or variables. The object y is defined by two subject-relation pairs and is used for defining the object x. In this architecture, there is a forward

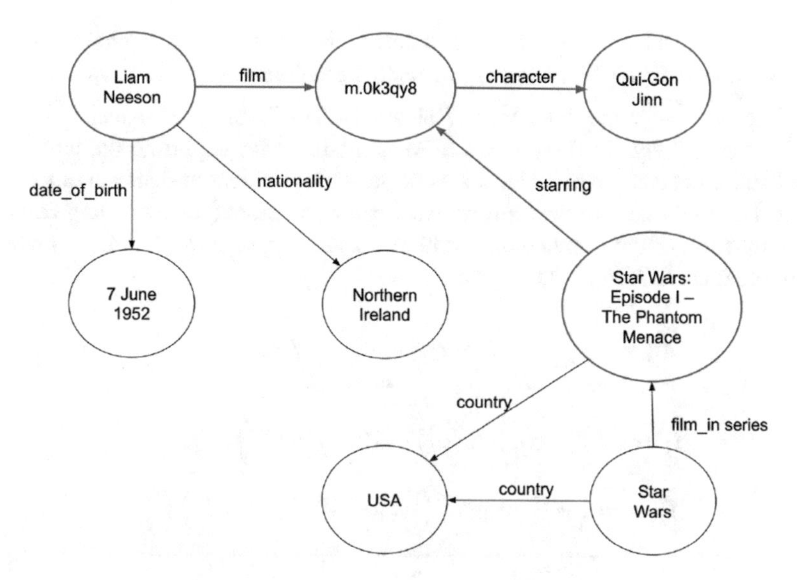

Fig. 6.21 The subgraph of Freebase and a candidate query generated for the mentioned question (Connections and entities of the candidate query are shown with red color)

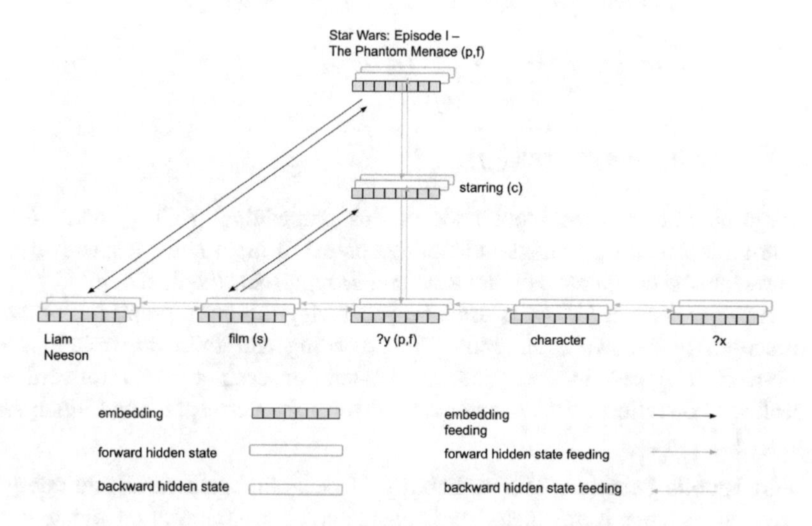

Fig. 6.22 Tree-based encoder

path, named forward directions (the opposite path of forward directions is named backward directions), beginning from each node or relation to the answer node. Each node or relation has three kinds of context information, namely, (1) preceding context, (2) following context, and (3) sibling context. The preceding context of node a is the node that appears next to it, while the following context of node a is the node which node a flows to. The sibling context of node a is node

b which has a common following context with node a. According to the example provided in Fig. 6.22, preceding context, sibling context, and following context of node $\xrightarrow{\text{starring}}$ are "Star Wars: Episode I – The Phantom Menace," $\xrightarrow{\text{film}}$, and $?y$, respectively. The standard LSTM is modified for capturing the embedding of sibling context and hidden states of preceding context besides the embedding of the input node. In the case of existing several sibling or preceding contexts, summation of their embedding or hidden state is passed to LSTM. Formulation of modified LSTM is as follows:

$$
\begin{aligned}
\vec{\tilde{h}}_j &= \sum_{a_k \in PC(a_j)} \vec{h}_k, z_j = \sum_{a_{k'} \in SC(j)} a_{k'}, \\
i_j &= \sigma\left(W^{(i)} a_j + U^{(i)} \hat{h}_j + V^{(i)} z_j + b^{(i)}\right), \\
f_{jk} &= \sigma\left(W^{(f)} a_j + U^{(f)} \vec{h}_k + V^{(f)} z_j + b^{(f)}\right), \\
o_j &= \sigma\left(W^{(0)} a_j + U^{(0)} \vec{h}_j + V^{(0)} z_j + b^{(0)}\right), \\
u_j &= \tanh\left(W^{(u)} a_j + U^{(u)} \tilde{h}_j + V^{(u)} z_j + b^{(u)}\right), \\
c_j &= i_j \odot u_j + \sum_{a_k \in PC(a_j)} f_{jk} \odot c_k, \\
\vec{h}_j &= o_j \odot \tanh\left(c_j\right),
\end{aligned}
\tag{6.28}
$$

where a_j indicates the input node or link embedding; $PC(a_j)$ and $SC(a_j)$ return the preceding context and sibling context of input node, respectively; u_j represents the candidate cell memory; and c_j represents the final cell.

Another distinct LSTM is used for processing the input graph in backward direction. In backward direction, the preceding and following relations are reversed. The two hidden states, generated for each node in forward and backward directions, are concatenated to form the corresponding hidden state $h_j = \left\{\vec{h}_j; \overleftarrow{h_j}\right\}$.

- **Mixed-mode decoder:** The probability of decoding each candidate query to question sentence is calculated and considered as matching of candidate query and question. The decoding model includes two generating and referring modes and calculates the probability of decoding question q having the candidate query a as follows:

$$P(q \mid a) = \prod_{t=1} P(q_t \mid q_{<t}, a)$$
$$= \prod_{t=1}^{n} [P(gen \mid q_{<t}) P_{\text{gen}} (q_t \mid q_{<t}, a) + P(\text{ref} \mid q_{<t}) P_{\text{ref}} (q_t \mid q_{<t}, a)] \tag{6.29}$$

where n is the length of question q, $q_{<t}$ indicates $q_0, \cdots, q_{t-1}$, and gen and ref are the generating and referring modes. Two distinct probabilities are calculated for each token, and they are combined by using a gate mechanism.

The generating mode is used for recognizing the non-informative words which are defined as words not helping to identify relations of the given question. For example, the word "who" in question "Who is Liam Neeson's youngest child?" cannot help to identify the relation $\xrightarrow{\text{children}}$. The generating mode works by modeling these words. In the generating probability, the probability of each token $P_{\text{gen}}(q_t \mid q_{<t}, a)$ is estimated by using a target-side vocabulary V_g and calculating the context representation c_t of the query a using encoder states H as follows:

$$c_t = \sum_{\tau=1}^{T_s} \alpha_{\tau t} h_\tau, \quad \alpha_{\tau t} = \frac{e^{\eta(s_{t-1}, h_\tau)}}{\sum_{\tau'} e^{\eta(s_{t-1}, h_{\tau'}})} \tag{6.30}$$

where α is the attention weight and η is a MLP for calculating the importance of the current hidden state.

$$P_{\text{gen}} (q_t \mid q_{<t}, a) = P_{\text{gen}} (q_t \mid q_{t-1}, \mathbf{s}_{t-1}, c_t)$$
$$= \frac{\exp\left(q_t^\top W_g [q_{t-1}; s_{t-1}; c_t]\right)}{\sum_{v_i \in V_g} \exp\left(v_i^\top \mathbf{W}_g [q_{t-1}; \mathbf{s}_{t-1}; c_t]\right)} \tag{6.31}$$

where s_{t-1} represents the decoder hidden state, W_g indicates the transformation matrix, and when $q_{t-1} \notin V_g$, it is considered a specific symbol $\_UNK\_ \in V_g$.

Generating mode cannot capture the semantic similarity of the question and query alone as (1) the limited size of the target-side vocabulary is not enough for capturing the correlation of question and query on the surface level and (2) the target-side vocabulary cannot cover the different expressions of entities or relations. The mentioned limitations of the generating mode are solved by using referring mode. Referring mode identifies the informative words that are not rare in questions and help the model to recognize relations. The informative words can be a name or an alias/paraphrase. The name "Neeson" and alias/paraphrase "job" referring to $\xrightarrow{\text{profession}}$ are examples of the informative words. A language model which is trained on entity/relation of the queries by using the Web data for extracting aliases/paraphrases is used. The backoff smoothing is applied to a trigram language model and trained offline for predicting each entity and relation. Given the query a, the referring probability is calculated as:

$$\begin{aligned} P_{\text{ref}}\left(q_t \mid q_{<t}, a\right) &= P_{\text{ref}}\left(q_t \mid q_{t-2}, q_{t-1}, \boldsymbol{\alpha}_t, \boldsymbol{L}\right) \\ &= \sum_{j}^{a_j \in a \cap (E \cup R)} \alpha_j P_{L_j}\left(q_t \mid q_{t-2}, q_{t-1}\right) \end{aligned} \tag{6.32}$$

where $\boldsymbol{\alpha}_t$ is the alignment probability, $\boldsymbol{L}$ is a set of language models trained for entities and relations, and $a_j \in a \cap (E \cup R)$ is the corresponding language model of a_j. A sequence of infrequent terms that represent an entity/relation is captured by a referring node.

The gate is used for indicating whether the next word points to an entity/relation or should be utilized in smoothing based on the previous observations. For example, when a question begins with "who is," the probability of observing a person's name after this sequence is high. The probabilities of referring and generating modes are generated by using an MLP as follows:

$$(P(gen), P(ref)) = m(s_{t-1}) \tag{6.33}$$

where m represents the MLP.

The decoder state or the gate, indicated by s_t, specifies which word should be considered in generating mode, and it is calculated as follows:

$$\begin{aligned} s_t &= LSTM\left(s_{t-1}, \left[\boldsymbol{q}_t; c_t; \boldsymbol{l}_t\right]\right) \\ \boldsymbol{l}_t &= \sum_{j}^{a_j \in a \cap (E \cup R)} \frac{P_{L_j}\left(q_t \mid q_{t-2}, q_{t-1}\right)}{Z} \boldsymbol{h}_j \end{aligned} \tag{6.34}$$

where Z represents the normalization parameter.

The model proposed by Hao et al. (2019) selects candidate DAGs from KB according to the entities of the question sentence. The encoded representations of the question sentence and the candidate DAGs are compared to find the most relevant DAG. In QAmp, the complex question is separated into a set of simple questions in a way that the answer of a question is used as a variable in the next question. Each of these questions is represented as a hub. QAmp selects a set of entities and predicates and assigns a confidence score to them. Then the confidence scores over the graph are aggregated by iteratively traversing the initial graph for concluding the answer. The SQG model assumes each question as a walk in KG. It forms a subgraph for each question and extracts the candidate walks from KG. Each candidate walk is converted into a SPARQL query. A tree-LSTM model is used for generating the representation of each question and candidate walks. Latent representations of the question and candidate walks are compared by computing a similarity score, and the most similar SPARQL query is selected. In the model proposed by Zhu et al. (2020), candidate queries for answering the question are selected from KG and encoded using tree-based encoder. The probability of decoding each candidate query to the question sentence is calculated, and the most probable query is selected.

6.5 Summary

We started this chapter with a short introduction of KBQA. In Sect. 6.2, we studied the traditional models proposed for KBQA. The questions are divided into two types including the simple questions and complex questions. We studied the models proposed for KBQA based on the type of the questions in two separate sections, namely, simple QA (Sect. 6.3) and complex QA (Sect. 6.4), respectively.

References

Bahdanau, D., Cho, K., & Bengio, Y. (2014). Neural machine translation by jointly learning to align and translate. arXiv preprint arXiv:1409.0473

Bao, J., Duan, N., Zhou, M., & Zhao, T. (2014). Knowledge-based question answering as machine translation. In *Proceedings of the 52nd Annual Meeting of the Association for Computational Linguistics (Volume 1: Long Papers)*, Baltimore, Maryland (pp. 967–976). Association for Computational Linguistics. https://doi.org/10.3115/v1/P14-1091

Bird, S., & Loper, E. (2004). NLTK: The natural language toolkit. In *Proceedings of the ACL Interactive Poster and Demonstration Sessions*, Barcelona, Spain (pp. 214–217). Association for Computational Linguistics. https://www.aclweb.org/anthology/P04-3031

Bordes, A., Usunier, N., Garcia-Durán, A., Weston, J., & Yakhnenko, O. (2013). Translating embeddings for modeling multi-relational data. In *Proceedings of the 26th International Conference on Neural Information Processing Systems - Volume 2, NIPS'13*, Red Hook, NY, USA (pp. 2787–2795). Curran Associates.

Brown, P. F., Della Pietra, S. A., Della Pietra, V. J., & Mercer, R. L. (1993). The mathematics of statistical machine translation: Parameter estimation. *Computational Linguistics, 19*(2), 263–311. https://www.aclweb.org/anthology/J93-2003

Finkel, J. R., Grenager, T., & Manning, C. D. (2005). Incorporating non-local information into information extraction systems by Gibbs sampling. In *Proceedings of the 43rd Annual Meeting of the Association for Computational Linguistics (ACL'05)*, Ann Arbor, Michigan (pp. 363–370). Association for Computational Linguistics. https://doi.org/10.3115/1219840.1219885

Hao, Z., Wu, B., Wen, W., & Cai, R. (2019). A subgraph-representation-based method for answering complex questions over knowledge bases. *Neural Networks, 119*, 57–65.

Golub, D., & He, X. (2016). Character-level question answering with attention. In *Proceedings of the 2016 Conference on Empirical Methods in Natural Language Processing*, Austin, Texas (pp. 1598–1607). Association for Computational Linguistics. https://doi.org/10.18653/v1/D16-1166

Huang, X., Zhang, J., Li, D., & Li, P. (2019). Knowledge graph embedding based question answering. In *Proceedings of the Twelfth ACM International Conference on Web Search and Data Mining, WSDM '19*, New York, NY, USA (pp. 105–113). Association for Computing Machinery. https://doi.org/10.1145/3289600.3290956

Lin, Y., Liu, Z., Sun, M., Liu, Y., & Zhu, X. (2015). Learning entity and relation embeddings for knowledge graph completion. In *Proceedings of the Twenty-Ninth AAAI Conference on Artificial Intelligence, AAAI'15* (pp. 2181–2187). AAAI Press. ISBN: 0262511290.

Mohammed, S., Shi, P., & Lin, J. (2018). Strong baselines for simple question answering over knowledge graphs with and without neural networks. In *Proceedings of the 2018 Conference of the North American Chapter of the Association for Computational Linguistics: Human Language Technologies, Volume 2 (Short Papers)*, New Orleans, Louisiana (pp. 291–296). Association for Computational Linguistics. https://doi.org/10.18653/v1/N18-2047.

Och, F. J. (2003). Minimum error rate training in statistical machine translation. In *Proceedings of the 41st Annual Meeting of the Association for Computational Linguistics*, Sapporo, Japan (pp. 160–167). Association for Computational Linguistics. https://doi.org/10.3115/1075096.1075117

Och, F. J., & Ney, H. (2003). A systematic comparison of various statistical alignment models. *Computational Linguistics, 29*(1), 19–51. https://doi.org/10.1162/089120103321337421

Orr, D., Subramanya, A., Gabrilovich, E., & Ringgaard, M. (2011). Billion clues in 800 million documents: A web research corpus annotated with freebase concepts. *Google Research Blog, 11*.

Sorokin, D., & Gurevych, I. (2017). End-to-end representation learning for question answering with weak supervision. In M. Dragoni, M. Solanki, & E. Blomqvist (Eds.), *Semantic Web Challenges*. Springer.

Unger, C., Bühmann, L., Lehmann, J., Ngonga Ngomo, A.-C., Gerber, D., & Cimiano, P. (2012). Template-based question answering over rdf data. In *Proceedings of the 21st International Conference on World Wide Web, WWW '12*, New York, NY, USA (pp. 639–648). Association for Computing Machinery. ISBN 9781450312295. https://doi.org/10.1145/2187836.2187923

Vakulenko, S., Fernandez Garcia, J. D., Polleres, A., de Rijke, M., & Cochez, M. (2019). Message passing for complex question answering over knowledge graphs. In *Proceedings of the 28th ACM International Conference on Information and Knowledge Management* (pp. 1431–1440).

Yang, Y., & Chang, M. W. (2015). S-MART: Novel tree-based structured learning algorithms applied to tweet entity linking. In *Proceedings of the 53rd Annual Meeting of the Association for Computational Linguistics and the 7th International Joint Conference on Natural Language Processing (Volume 1: Long Papers)*, Beijing, China (pp. 504–513). Association for Computational Linguistics. https://doi.org/10.3115/v1/P15-1049

Yao, X., & Van Durme, B. (2014). Information extraction over structured data: Question answering with Freebase. In *Proceedings of the 52nd Annual Meeting of the Association for Computational Linguistics (Volume 1: Long Papers)*, Baltimore, Maryland (pp. 956–966). Association for Computational Linguistics. https://doi.org/10.3115/v1/P14-1090

Yin, J., Jiang, X., Lu, Z., Shang, L., Li, H., & Li, X. (2016a). Neural generative question answering. In *Proceedings of the Workshop on Human-Computer Question Answering*, San Diego, California (pp. 36–42). Association for Computational Linguistics. https://doi.org/10.18653/v1/W16-0106

Yin, W., Yu, M., Xiang, B., Zhou, B., & Schütze, H. (2016b). Simple question answering by attentive convolutional neural network. In *Proceedings of COLING 2016, the 26th International Conference on Computational Linguistics: Technical Papers*, Osaka, Japan (pp. 1746–1756). The COLING 2016 Organizing Committee. https://www.aclweb.org/anthology/C16-1164

Zafar, H., Napolitano, G., & Lehmann, J. (2018). Formal query generation for question answering over knowledge bases. In *European Semantic Web Conference* (pp. 714–728). Springer.

Zhu, S., Cheng, X., & Su, S. (2020). Knowledge-based question answering by tree-to-sequence learning. *Neurocomputing, 372*, 64–72. ISSN 0925-2312. https://doi.org/10.1016/j.neucom.2019.09.003

Chapter 7
KBQA Enhanced with Textual Data

Abstract This chapter studies the KBQA models enhanced with textual data. These models benefit from both KG and text for answering the questions. We have studied the QA models which rely on only text or only KGs as source of knowledge in the previous chapters. Regarding the limitations of the models which only rely on one source of knowledge, we will see how using a combination of text and KG can help the models for leveraging their performance.

7.1 Introduction

In addition to the KBQA and TextQA, there is another type of QA which uses the information from textual data in KBQA. The textual data can be combined in different stages of a KBQA model. The textual information can be used in refining the answers retrieved from KG. Also, it can be used in identifying the potential entities of the KG or constructing a subset of KG.

By analyzing the performance of the KBQA models, it can be seen that questions which cover a wide range of entities in the KG are among the most difficult questions for KBQA models. In this case, using textual data can help the models to identify more relevant nodes and eliminate the non-relevant nodes.

In the following section, we will explain the architecture of some well-known models which use textual data in KBQA models.

7.2 Study of Models

Text2KB (Savenkov & Agichtein, 2016) is inspired by the idea of enriching the KBQA using textual data. The architecture of this model includes external textual data sources including CQA, collection of documents, and Web search that are used to enhance the information extraction process.

Utilizing the textual data helps the model in the challenging tasks including (1) identifying entities, (2) scoring predicates, and (3) ranking the candidate answers.

S. Momtazi, Z. Abbasiantaeb, *Question Answering over Text and Knowledge Base*,
https://doi.org/10.1007/978-3-031-16552-8_7

In the following, we will explain how the textual data is used in different parts of the existing KBQA models to improve them.

- **Web search results for KBQA:** The question sentence is used for querying the search engines. The first ten most relevant snippets returned by Web search along with the source documents are returned. In the next step, the knowledge graph entity mentions are discovered in both the snippets and corresponding documents by utilizing the Aqqu entity linking module.

  Identifying the entities of question is a challenging task due to the limited context of the question sentence, misspelling in entity names, and using an uncommon version of entity names. The documents returned by Web search help to mitigate the mentioned challenges as they include several mentions of the same entity and a larger context. The topical entities of the input question are expanded by using the retrieved snippets. The most frequent entities occurring in the snippets that are similar to the question words are selected and added to topical entities.

  For calculating the similarity of a name from snippet and a question term q_t, the Jaro-Winkler distance model is used as follows:

$$\max_{e_t \in M \setminus \text{Stop}, q_t \in Q \setminus Stop} 1 - \text{dist}(e_t, q_t) \geq 0.8 \tag{7.1}$$

  where e_t is a token of entity M except the stop words. An entity is selected according to the above function if at least one of its tokens is similar to a question term. Features are extracted for each answer candidate as follows:

  1. IDF of words and entities,
  2. Two TF-IDF vectors generated for each snippet and document, one for entities (e_{s_i}, e_{d_i}) and other for lower-cased terms (t_{s_i}, t_{d_i}),
  3. Vector representation of all documents and snippets generated by merging the vector representation of the documents and the snippets $(t_{\cup s_i}, t_{\cup d_i}, e_{\cup s_i}, e_{\cup d_i})$,
  4. TF-IDF vector representation of answer candidate a_j based on terms t_{a_j} and entities e_{a_j},
  5. Average and maximum value of the four cosine similarities: $\cos\left(t_{s_i}, t_{a_j}\right)$, $\cos\left(t_{d_i}, t_{a_j}\right)$, $\cos\left(e_{s_i}, e_{a_j}\right)$, and $\cos\left(e_{d_i}, e_{a_j}\right)$,
  6. Four different cosine similarities: $\cos\left(e_{\cup s_i}, e_{t_j}\right)$, $\cos\left(e_{\cup s_i}, e_{a_j}\right)$, $\cos\left(t_{\cup d_i}, t_{a_j}\right)$, and $\cos\left(e_{\cup d_i}, e_{a_j}\right)$.

- **CQA data for matching questions to predicates:** This step is inspired by the work of Savenkov et al. (2015) which uses distant supervision for extracting knowledge from CQA dataset to complete the knowledge base. A series of weakly labeled question and answer pairs of Yahoo! WebScope L dataset is used for calculating the association of question words and predicates to expand the lexicon. A sample of this dataset with 4.4M questions is utilized.

  The entities of the question and the answer sentences are linked by using an entity linker model which discovers mentions from Freebase entities. A distant

Table 7.1 Example of PMI score for term-predicate pairs calculated by the distance supervision method on a CQA corpus (Savenkov & Agichtein, 2016)

Term	Predicate	PMI
Born	people.person.date_of_birth	3.67
	people.person.date_of_death	2.73
	location.location.people_born_here	1.60
Kill	people.deceased_person.cause_of_death	1.70
	book.book.characters	1.55
Currency	location.country.currency_formerly_used	5.55
	location.country.currency_used	3.54
School	education.school.school_district	4.14
	people.education.institution	1.70
	sports.school_sports_team.school	1.69
Win	sports.sports_team.championships	4.11
	sports.sports_league.championship	3.79

supervision model is utilized to predict the label for matching the entities from question sentence and the answer sentence. The association of question words and predicates is learned using these labels with calculating the point-wise mutual information (PMI) score among the question words and predicates. An example of calculating the PMI metric for question words and predicates is shown in Table 7.1.

In Text2KB, the evaluation of candidate answer predicates is performed with PMI score of question words and predicates. Three features are obtained for each candidate answer by using the minimum, maximum, and average of these PMI values. In addition, predicate encoding is obtained by calculating the weighted average of the Word2Vec representation from predicate's PMI table and compared with question words by cosine similarity. The minimum, maximum, and average of these values are considered as features. At the end, the average of question word embeddings is calculated, and its cosine similarity with the predicate vector is calculated.

- **Estimating entity associations:** The text data can be used for estimating the association of entities because the entities occurring in the same context are related to each other. For instance, having a passage that includes the name of two entities is a good resource for extracting the relation of these entities as with a high probability it is explaining their relationship.

  The contextual information of these two entities can be represented as an extra edge that connects the corresponding edges in the knowledge graph. The ClueWeb12 corpus and the entity annotations of Freebase are utilized in this step. The number of distinct words in the context of a mention of entity pairs that occur within the previous and next 200 characters is counted. In this unigram language model, the words between two entities and the words within a small window of mentions are considered. After obtaining the unigram probabilities, the language model probability of each entity pair e_1 and e_2 given the question Q is calculated as follows:

$$p(Q \mid e_1, e_2) = \prod_{t \in Q} p(t \mid e_1, e_2) \tag{7.2}$$

The maximum, average, and minimum language model score of all entity pairs are considered as features. For mitigating the sparsity problem, we can use the word embeddings. In the case of using the word embeddings, each entity pair is represented with a weighted average of words. Then, the cosine similarity of entity pair embedding and the question words is calculated. The minimum, average, and maximum values of the cosine similarities are considered as features.

- **Internal text data to enrich entity representation:** Most of the knowledge bases contain text data. This text data can be used for enriching the entity representation. For example, Freebase includes a paragraph extracted from Wikipedia for describing each entity. The vector representation of each entity description is generated with tokens and entities.

Hybrid QA is proposed by Xu et al. (2016a) and is a hybrid QA system which tries to mitigate the limitations which KBQA faces due to the closed vocabulary of predicates. These limitations are listed below:

1. The coverage is limited as predicates are gathered by community,
2. Eliminating potential relevant semantic differences is probable,
3. The possibility of providing an imperfect answer despite having a logical form due to the incompleteness of knowledge graphs.

Hybrid QA uses both structured data from knowledge base and unstructured data from text to capture meaning of the input question. The architecture of hybrid QA in processing the input question "Who is the front man of the band that wrote Coffee and TV?" is shown in Fig. 7.1.

The hybrid QA model works in two major steps. In the first step, entity linking and relation extraction are performed to predict the entities and relations. The entities that are relevant to the input question are extracted from the knowledge graph by the entity linking module, and the relations among question words or entities are predicted by two distinct models executed over knowledge graph and textual data. A neural network-based model is proposed for predicting the predicates from a knowledge graph given the relational phrases. The textual relations that most likely explain the phrases are discovered by running a paraphrase model.

In the second step, the optimal configuration is found by applying a joint interface on the output of entity linking and relation extraction.

The example question processed in Fig. 7.1 is a complex question that requires multiple facts or constraints from the knowledge graph to be answered. The constraints are (1) the intended person is the front man of a band and (2) the band wrote a song named Coffee and TV. By utilizing the syntactic rules proposed by (Xu

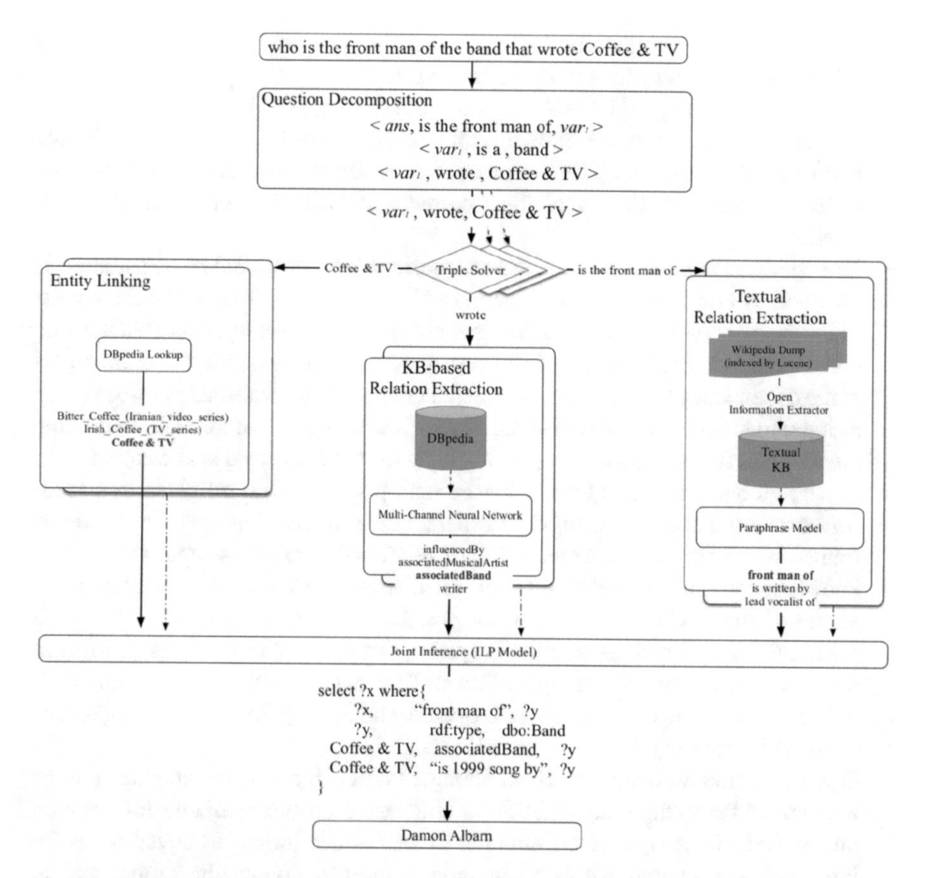

Fig. 7.1 The architecture of the hybrid QA model (Xu et al., 2016a) in processing the question: "Who is the front man of the band that wrote Coffee and TV?" ("©1963–2022 ACL, reprinted with permission")

et al., 2016b), the question is parsed into three simple questions represented by the following triples:

(1) $< ans, \textit{is the front man of}, var_1 >$
(2) $< var_1, \textit{is a}, \textit{band} >$
(3) $< var_1, \textit{wrote}, \textit{Coffee \& TV} >$

The hybrid QA model follows an information extraction-based approach including entity linking and relation extraction over both the textual relations and the knowledge graph predicates. The overall architecture of hybrid QA is explained in the following steps.

- **Entity linking:** A recall-oriented method is preferred to a precision-oriented method because by extracting all of the probable entities, we can guarantee having the correct entities in the initial step and the correctness of the retrieved entities will be determined in the following steps.

DBpedia Lookup and S-MART (Yang and Chang, 2015) methods are applied for retrieving the top 10 entities from source knowledge graphs, including Freebase and DBpedia. The entities extracted by the mentioned methods are called candidate entities, and their correctness will be checked in the joint interface step.

- **Knowledge graph-based relation extraction:** The relational representation is obtained by employing a multi-channel CNN (Xu et al., 2016b). The three parts of the facts including subject, relational phrase, and object are concatenated and considered as a sentence. The shortest path connecting the subject and the object in the dependency tree is passed to the first channel. The relational phrase is given as input to the second channel. A CNN with a window size of three for extracting the tri-gram features by a max-pooling layer on top is utilized as a channel.

  The global features extracted by the max-pooling layer are given to a feed-forward neural network as input to generate the semantic features. The model is trained by using a dataset consisting of relational phrases as input and related knowledge graph predicates as output. A distribution vector is generated in the output of the model over a set of all possible predicates. Two distinct models are trained upon Freebase and the DBpedia, since the set of predicates differs in the mentioned knowledge graphs. The PATTY dataset with 127,811 samples is utilized for training the model over DBpedia, and 3,022 pairs from WebQuestions are used for training the model over Freebase.

- **Open relation extraction:** Even though knowledge graphs include a large number of facts, their information is still less than the available information on the Web. To mitigate the limitation of knowledge graphs in covering all the information, a paraphrase-based method is used to predict the proper textual relations given the relational phrases as input.

  A textual knowledge repository is constructed by extracting $< argument_1$, *relation*, $argument_2 >$ triples from English Wikipedia. The triples are extracted by applying the information extractor model proposed by Angeli et al. (2015).

  In the paraphrasing task, a relational phrase rp is given as input, and the most similar textual relation is selected from the candidate textual relations $TR = \{tr_1, tr_3, \cdots, tr_{|TR|}\}$. Various paraphrasing methods based on dynamic pooling and recursive auto-encoders (Socher et al., 2011) are used in this model. These paraphrasing models extract feature vectors for phrases constructed in the syntactic tree.

  The similarity of two sentences is calculated in the phrase and word levels based on the extracted features. A similarity matrix is generated for phrase-label and word-level similarity of the two input sentences.

  Due to the variable length of sentences, the length of the similarity matrix is not fixed. To obtain a fixed-sized vector representation, a dynamic pooling layer is applied on the similarity matrix, and the pooled representation is passed to a classifier C_p.

The recursive auto-encoders, which are pre-trained on the two sections of Gigaword corpus (NYT and AP sections), are used. The C_p classifier is trained on a monolingual corpus named PARALEX (Fader et al., 2013). The PARALEX corpus includes 18 million question phrase pairs which the questions having the same meaning are tagged.

- **Joint inference:** The optimal configuration of entity and relational phrases is concluded in this step. At the first step, the coherence of the predicate and entity is examined. When a relational phrase rp is related to a knowledge graph predicate kr, the semantic types of the entities are matched with the expectation of the predicate.

  The subject entity's type is extracted from the knowledge graph schema, and all of the entities placed in the subject position of the corresponding predicate are extracted to check if at least one of them has the same type as the subject entity's type. The initial value of $Coh_{e,kr}$ is set to zero, and if there exists an entity with the same type, it changes to $Coh_{e,kr} = 1$.

  Like the first step, the coherence of the textual relation and entity is matched. As the textual relations lack the schema, the types of the entities connected to the textual relation tr as subject or object are considered as types of tr. If the type of the entity e exists among the retrieved types, the entity e is compatible with tr, and the value of $Coh_{e,tr}$ is set to one ($Coh_{e,tr} = 1$).

  In the third step, the compatibility of the knowledge graph predicate kr and textual relation tr is checked because a relational phrase can be mapped to both the knowledge graph predicate and the textual relation. At first, kr and tr are checked to determine if they have similar argument expectations. In the case of having similar argument expectations, a pre-trained multi-channel CNN is utilized for modeling $Coh_{kr,tr}$. The value of $Coh_{kr,tr}$ is in range of (0, 1) that determines the probability of mapping tr to kr and $Coh_{kr,tr} = -1$ when kr and tr have no similar arguments.

  After completing the above three steps, their outputs are aggregated by forming an integer linear program (ILP) formulation. By completing the above steps, the following objective function must be maximized:

$$\max \quad \alpha \times \text{conf}^e + \beta \times \text{con}\, f^r + \delta \times \text{conf}^{er} \tag{7.3}$$

where conf^e and $\text{con}\, f^r$ represent the score of entity linking and relation extraction, respectively. α, β, and δ are weighting parameters. conf^e is calculated as follows:

$$\text{conf}^e = \sum_{d} \sum_{ep \in d, e \in C_e(ep)} w_{ep,e} Y_{ep,e} \tag{7.4}$$

where ungrounded triple is represented with d, $C_e(ep)$ returns the set of candidate entities of the input entity phrase (ep), $Y_{ep,e}$ indicates whether the given entity phrase ep maps the given entity e or not, and $w_{ep,e}$ represents the entity linking score. The value of conf^r is calculated based on the following equation:

$$\text{conf}^r = \sum_d \sum_{rp \in d, kr \in C_{kr}(rp)} q_{rp,kr} Z_{rp,kr} + \sum_d \sum_{rp \in d, tr \in C_{tr}(rp)} v_{rp,tr} W_{rp,tr} \tag{7.5}$$

where $C_{kr}(rp)$ and $C_{tr}(rp)$ return a set of candidate knowledge graph predicates and textual relations given the input relational phrase rp, respectively. The relational phrase rp is mapped to knowledge graph and textual relations with $q_{rp,kr}$ and $v_{rp,tr}$ scores, respectively. The mapping of relational phrase rp with knowledge graph and textual relations is denoted by $Z_{rp,kr}$ and $W_{rp,tr}$, respectively. Finally, the coherence of the candidate entities and relations is defined as:

$$\begin{aligned} \text{conf}^{er} &= \sum_d \sum_e \sum_{kr} o_{e,kr} \text{Coh}_{e,kr} \\ &+ \sum_d \sum_e \sum_{tr} o_{e,tr} Coh_{e,tr} \\ &+ \sum_d \sum_{kr} \sum_{tr} o_{kr,tr} Coh_{kr,tr} \end{aligned} \tag{7.6}$$

where $o_{a,b}$ denotes the coherence score between a and b and $Coh_{a,b}$ denotes whether both the semantic components a and b are chosen or not.

The following constraints that are defined for solving the ILP problem are defined below. The first constraint states that each entity phrase must be mapped to just one entity as:

$$\forall d, \forall e \in C_e(ep), \quad \sum_{ep \in d, e \in C_e(ep)} Y_{ep,e} \leq 1 \tag{7.7}$$

The second constraint states that each relational phrase must be mapped to at most one knowledge graph and textual relations as:

$$\forall d, \forall kr \in C_{kr}(rp), \quad \sum_{rp \in d, kr \in C_{kr}(rp)} Z_{rp,kr} \leq 1 \tag{7.8}$$

$$\forall d, \forall tr \in C_{tr}(rp), \quad \sum_{rp \in d, tr \in C_{tr}(rp)} Z_{rp,tr} \leq 1 \tag{7.9}$$

The following constraints are defined to ensure that the value of variables $Coh_{e,kr}$, $Coh_{e,tr}$, and $Coh_{kr,tr}$ equals one just when the value of corresponding Y and Z variables is one as follows:

$$\forall d, \forall e \in C_e(ep), \forall kr \in C_{kr}(rp), \forall tr \in C_{tr}(rp) \tag{7.10}$$

$$Coh_{e,kr} \leq Y_{ep,e} \quad Coh_{e,kr} \leq Z_{rp,kr} \quad Y_{ep,e} + Z_{rp,kr} \leq 1 + Coh_{e,kr} \tag{7.11}$$

$$Coh_{e,tr} \leq Y_{ep,e} \quad Coh_{e,tr} \leq W_{rp,tr} \quad Y_{ep,e} + W_{rp,tr} \leq 1 + Coh_{e,tr} \tag{7.12}$$

$$Coh_{kr,tr} \leq Z_{rp,kr} \quad Coh_{kr,tr} \leq W_{rp,tr} \quad Z_{rp,kr} + W_{rp,tr} \leq 1 + Coh_{kr,tr} \tag{7.13}$$

The above ILP problem is solved by Gurobi.

GRAFT-Net (Sun et al., 2018) consumes the combined data from heterogeneous sources including textual data and knowledge base by constructing a question-specific subgraph. The overall architecture of GRAFT-Net is shown in Fig. 7.2. Answer of the questions is assumed to be an entity, and given a question $q = \{w_1, \cdots, w_{|q|}\}$, the corresponding entity answer is extracted from both the knowledge graph and document entities.

A subgraph $\mathcal{G}_q \subset \mathcal{G}$ which ensures the existence of the correct answer is extracted. The constructed subgraph is passed to the GRAFT-Net model to generate the node representations and classify the nodes as being the answer or not.

A knowledge graph is formulated as a set of entities $\mathcal{V}$, edges $\mathcal{E}$, and relations $\mathcal{R}$. Each edge is represented as a triple (s, r, o) where $s \in \mathcal{V}$ and $o \in \mathcal{V}$ are subject and object entities connected with relation $r \in \mathcal{R}$. A text corpus is denoted as an array of documents $\mathcal{D} = \{d_1, \cdots, d_{|D|}\}$, where each document is represented as a sequence of words $d_i = (w_1, \cdots, w_{|d_i|})$. Given the input document d, an entity linking model outputs a set $\mathcal{L}$ consisting of links (v, d_p) where each link maps an entity v to a word d_p (at p-th index of the input document). Entity links associated with document d are denoted by $\mathcal{L}_d$. The link of an entity mention including several words of document d is constructed by combining links of all the words.

The subgraph $\mathcal{G}_q \subset \mathcal{G}$ of the input question q is constructed in a parallel process from both text and knowledge graph. Entities are extracted from the knowledge graph, and relevant documents are extracted from textual data. The entity links are used to combine the entities and the documents in a fully connected graph. The knowledge base and text retrieval approaches are explained below.

- **Knowledge base retrieval:** The relevant entities of the question sentence (called a set of seed entities S_q) are extracted by entity linking. The personalized PageRank method (Haveliwala, 2002) is executed over the seed entities to extract other entities that could be potential answers. The edges around the S_q are given weight based on their relatedness to the question, and the edges with similar

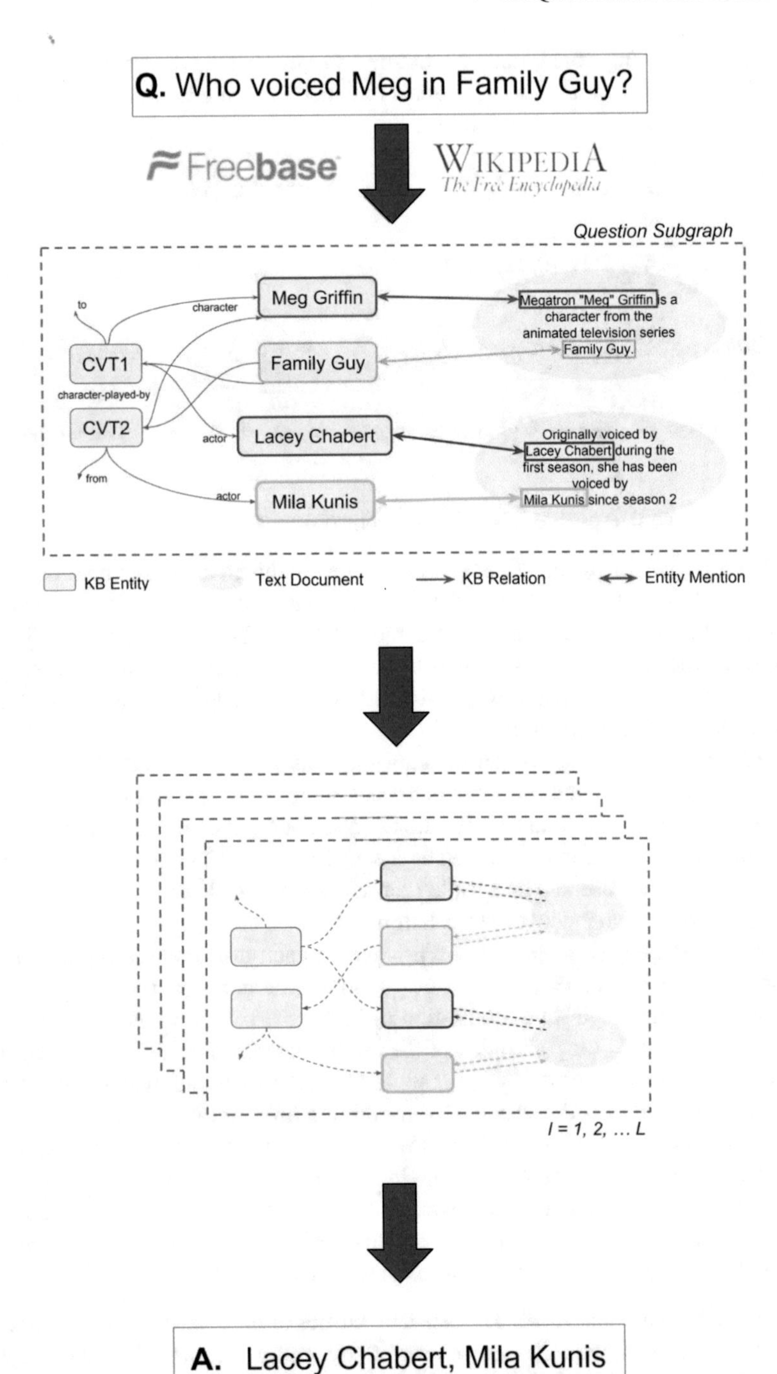

Fig. 7.2 **Up:** A graph is constructed from both knowledge base and text documents. **Down:** Node representations are generated and used for classifying them as the correct answer or not (Sun et al., 2018) ("©1963–2022 ACL, reprinted with permission")

type are weighted equally. In other words, the more relevant edges receive higher weight than the non-relevant edges. The vector representation of a relation ($v(r)$) is generated by averaging the word vectors of its surface form, and the vector representation of the question ($v(q)$) is generated by averaging the question word vectors. The edge weight is calculated by the cosine similarity between $v(r)$ and $v(q)$. The top E entities that have the highest personalized PageRank score are selected and added to $\mathcal{G}_q$ with the edges connecting them together.

- **Text retrieval:** Wikipedia is used as the text corpus, and each sentence of the Wikipedia articles is considered as a document. To retrieve the relevant text, two steps are required. In the first step, the weighted bag of words model from DrQA (Liang et al., 2017) is used for retrieving the relevant articles, and the first five articles are selected. In the second step, a Lucene index is applied over the articles for sentence retrieval, and the top relative sentences based on the question sentence are extracted as $d_1, \cdots, d_{|D|}$. The retrieved documents and the entities that are linked to them are appended to the subgraph $\mathcal{G}_q$.

  The subgraph constructed for question is denoted as $\mathcal{G}_q = (\mathcal{V}_q, \mathcal{E}_q, \mathcal{R}^+)$ where $\mathcal{V}$ includes all of the retrieved documents and the entities connected by edges from the knowledge graph $\mathcal{E}$ and the links between the entities and the documents. $\mathcal{V}_q$ and $\mathcal{E}_q$ are defined as:

$$\mathcal{V}_q = \{v_1, \ldots, v_E\} \cup \{d_1, \ldots, d_D\}$$
$$\mathcal{E}_q = \big\{(s, o, r) \in \mathcal{E} : s, o \in \mathcal{V}_q, r \in \mathcal{R}\big\} \cup \big\{\big(v, d_p, r_L\big) : \big(v, d_p\big) \in \mathcal{L}_d, d \in \mathcal{V}_q\big\} \tag{7.14}$$

  where all the edges in the subgraph are represented by $\mathcal{R}^+ = \mathcal{R} \cup \{r_L\}$ and r_L represents a special linking relation.

The answer of the question q is determined by labeling the nodes of $\mathcal{V}_q$. The label of the node v from $\mathcal{V}$ is equal to one ($y_v = 1$) when it belongs to the answer set $v \in \{a\}_q$, and the label of other nodes is equal to zero.

Some of the existing graph propagation-based methods (Schlichtkrull et al., 2018; Zhang et al., 2022) model the node representation and classify them. Most of these models initialize the node representations $h_v^{(0)}$ and update them through several layers (L layers) as below:

$$h_v^{(l)} = \phi\left(h_v^{(l-1)}, \sum_{v' \in N_r(v)} h_{v'}^{(l-1)}\right) \tag{7.15}$$

where ϕ indicates a neural layer and $N_r(v)$ returns neighbors of node v connected via an edge with type r.

Node representations are generated after completing the L layer (applying the neural layer for L times), and the obtained representations are used for other tasks like node classification and link prediction. This task is different from the mentioned graph-based classification problems in two ways: (1) the constructed graph includes two kinds of nodes (documents and entities) and is heterogeneous and (2) the node representation must be conditioned on the question sentence. The approach for mitigating the mentioned differences is explained in the following.

- **Node initialization:** The entity nodes are initialized randomly or by pre-trained knowledge graph embeddings. The first initialization of each entity node is represented with a fixed-size vector as $h_v^{(0)} = x_v \in \mathcal{R}^n$. As the documents have variable length, a variable length representation is considered for each document as $H_d^{(l)} \in \mathcal{R}^{|D| \times n}$.

  The representation of the document is initialized by using an LSTM neural network given the word embeddings $(w_1, \cdots, w_{|D|})$ as follows:

$$H_d^{(0)} = \text{LSTM}(w_1, w_2, \cdots) \tag{7.16}$$

  where the representation of the m-th word from document d in layer l is denoted as $H_{d,m}^l$.
- **Heterogeneous updates:** Two different update rules, as illustrated in Fig. 7.3, are used for updating the entity and document nodes.
  Entities. The entity nodes are updated by using a single-layer feed-forward neural network given four inputs as follows:

$$h_v^{(l)} = \text{FFN}\left(\left[\begin{array}{c} h_v^{(l-1)} \\ h_q^{(l-1)} \\ \sum_r \sum_{v' \in N_r(v)} \alpha_r^{v'} \psi_r\left(h_{v'}^{(l-1)}\right) \\ \sum_{(d,p) \in M(v)} H_{d,p}^{(l-1)} \end{array}\right]\right) \tag{7.17}$$

  where $h_v^{(l-1)}$ and $h_q^{(l-1)}$ represent entity and node representations and $M(v) = \{(d, p)\}$ returns a set of documents d and position of words p mentioning the entity v. The third term is created by summing the attentional transformed representation of the neighbor entities. The term $N_r(v)$ returns the neighbor entities of the input entity v, $\alpha_r^{v'}$ is attention weight, and ψ_r is transformation function. The following transformation matrix is used:

$$\psi_r\left(h_{v'}^{(l-1)}\right) = pr_{v'}^{(l-1)} \text{FFN}\left(x_r, h_{v'}^{(l-1)}\right) \tag{7.18}$$

  where $pr_{v'}^{(l-1)}$ is the PageRank score. The fourth term is summation over the representation of the words mentioning entity v.

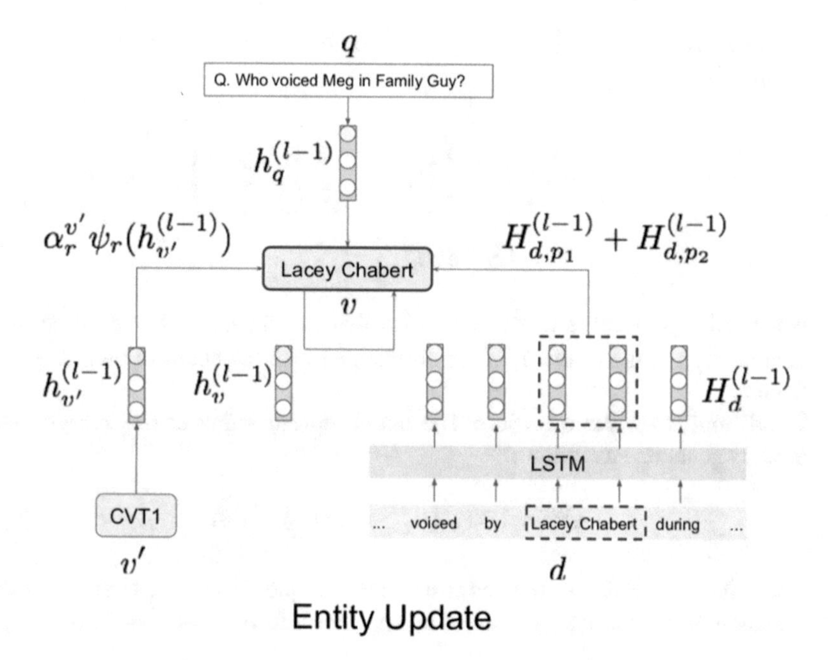

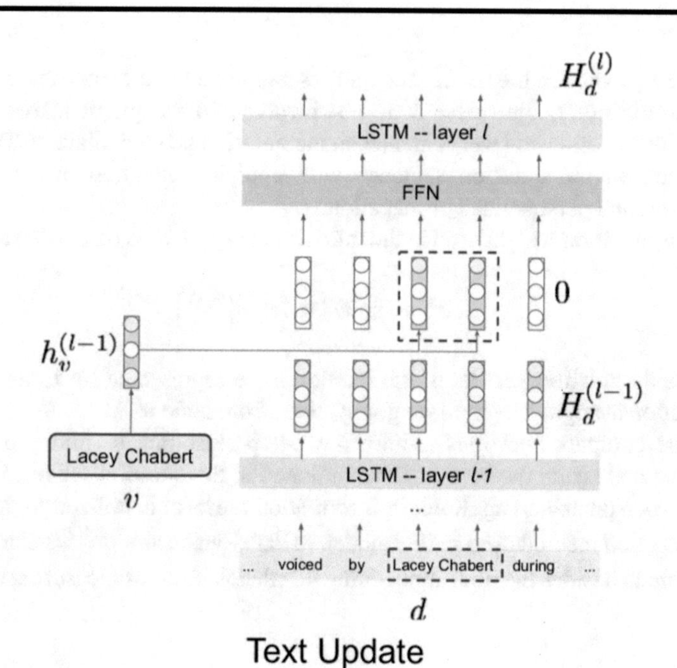

Fig. 7.3 Illustration of the update rules used for updating the entity and the document embeddings (Sun et al., 2018) ("©1963–2022 ACL, reprinted with permission")

Documents. The document nodes are updated according to the following equations:

$$\tilde{H}_{d,p}^{(l)} = \text{FFN}\left(H_{d,p}^{(l-1)}, \sum_{v \in L(d,p)} h_v^{(l-1)}\right)$$
$$H_d^{(l)} = \text{LSTM}(H_{d,p}^{(l)}) \tag{7.19}$$

where $L(d, p)$ returns a set of entities that are linked to the position p of document d. Value of the $h_v^{(l-1)}$ is normalized with the count of outgoing links from v.

- **Conditioning on the question:** The initial representation of the question sentence is generated as follows:

$$h_q^{(0)} = \text{LSTM}(w_1^q, \cdots, w_{|q|}^q)_{|q|} \in \mathcal{R}^n \tag{7.20}$$

where the last hidden state is considered as the question representation. Representation of the question is updated by applying a feed forward as follows:

$$h_q^{(l)} = \text{FFN}(\sum_{v \in S_q} h_v^{(l)}) \tag{7.21}$$

where S_q indicates the set of seed entities mentioned in the question sentence.

To this point, the question is not involved in the graph learner, and the dependency of the answer sentence on the question is not explained. The answer depends on the question sentence with applying attention mechanism over relations and personalized propagation.

The attention weight used in the third term of Eq. 7.17 is calculated as follows:

$$\alpha_r^{v'} = \text{softMax}\left(x_r^T h_q^{(l-1)}\right) \tag{7.22}$$

where the relation vectors of the relation r are represented by x_r and softMax operation is applied over all outgoing edges from node v'.

The complex questions require multi-hop reasoning to find a path to the answer node from the seed nodes mentioned in the question sentence. The idea of the personalized PageRank in information retrieval is utilized to develop the propagation method used in this model. A PageRank score $pr_v^{(l)}$ is calculated for each node besides the node embedding $h_v^{(l)}$ that is used for measuring the weight of path from seed entities to the answer node as follows:

$$pr_v^{(0)} = \begin{cases} \frac{1}{|S_q|} & \text{if} \quad v \in S_q \\ 0 & \text{o.w.} \end{cases} \tag{7.23}$$

$$pr_v^{(l)} = (1 - \lambda) pr_v^{(l-1)} + \lambda \sum_r \sum_{v' \in N_r(v)} \alpha_r^{v'} pr_{v'}^{(l-1)} \tag{7.24}$$

The attention score is also used in the above equation to ensure the nodes relevant to the question sentence are given a higher weight in calculating the weight of path. At first, only the seed entities have a PageRank score, and in the next layers, the PageRank score is propagated from the seed nodes to other nodes connected to the seed node. At the first layer ($l = 1$), the PageRank score is propagated to the nodes in one-hop distance from the seed entities.

- **Answer selection:** Representation of the entity nodes in the last layer is used for binary classification of the nodes. The nodes are classified as the correct answer of the question or an incorrect answer as follows:

$$\Pr\left(v \in \{a\}_q \mid \mathcal{G}_q, q\right) = \sigma\left(w^T h_v^{(L)} + b\right) \tag{7.25}$$

where W is the model parameter, b is the bias term, and σ indicates the sigmoid function.

The architecture of **KG-QA** framework (Tong et al., 2019) is shown in Fig. 7.4. The model includes two offline training and online serving steps. In the domain

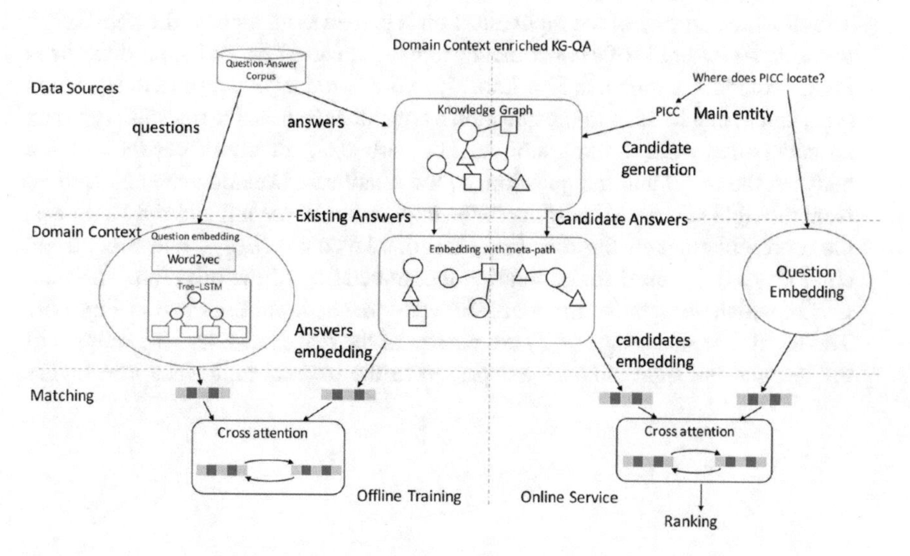

Fig. 7.4 Architecture of the KG-QA model (Tong et al., 2019) ("©Springer-Verlag GmbH Germany, part of Springer Nature, reprinted with permission")

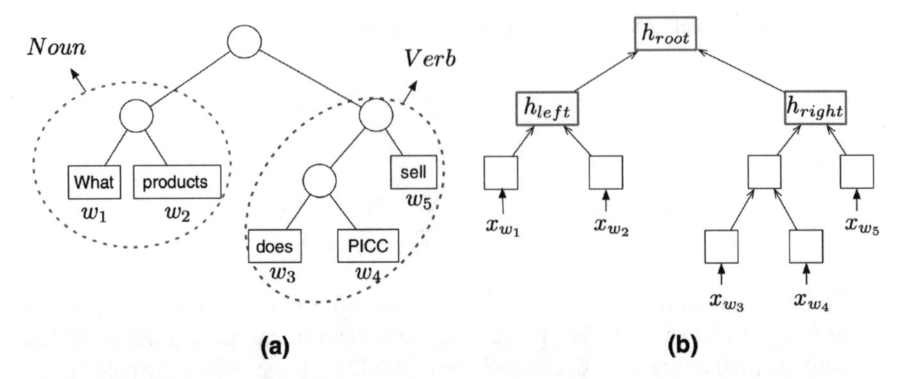

Fig. 7.5 Visualization of the constituency tree and the tree-LSTM for the example question (Tong et al., 2019) ("©Springer-Verlag GmbH Germany, part of Springer Nature, reprinted with permission"). (**a**) Constituency tree. (**b**) Tree LSTM architecture

context, the given question is processed by the question and answer representation modules. The model mainly works in three steps. In the first step, the question representation is generated. Then the candidate answers are selected from the KG and represented. In the third step, the similarity function S is used for calculating the similarity of the question and the candidate answers. We describe different components of this model in the following.

- **Input question representation:** The word embedding x_{w_i} is obtained by using the Word2Vec method on the related text. For example, for the insurance knowledge graph, the texts related to the insurance domain are used for generating the word embeddings. The question sentence is parsed using a tree-LSTM model. The constituency tree of the input question is passed as an input to the tree-LSTM network. An example of a constituency tree for the question "What products does PICC provide?" is shown in Fig. 7.5a. The word "product" represents the question type, and as can be seen in the question's constituency tree, the question type can be understood from the left subtree. The verb "sell" demonstrates the relation between the target and the question, so the relation information can be captured from the right subtree. According to the importance of the left and right subtrees, the representations of the root (h_{root}), root's left child (h_{left}), and root's right child (h_{right}) are used for generating the embedding of the question. The tree-LSTM which accepts the input constituency tree as input is shown in Fig. 7.5b. The word representations (x_{w_i}) are passed to the leaf nodes, and the outputs of the left and the right children are passed to the internal units. The tree-LSTM works according to the following equations:

$$
\begin{aligned}
i_j &= \sigma\left(W^{(i)}x_j + U_l^{(i)}h_{jl} + U_r^{(i)}h_{jr} + b^{(i)}\right)\\
f_{jk} &= \sigma\left(W^{(f)}x_j + U_{kl}^{(f)}h_{jl} + U_{kr}^{(f)}h_{jr} + b^{(f)}\right)\\
o_j &= \sigma\left(W^{(o)}x_j + U_l^{(o)}h_{jl} + U_r^{(o)}h_{jr} + b^{(o)}\right)\\
u_j &= \tanh\left(W^{(u)}x_j + U_l^{(u)}h_{jl} + U_r^{(u)}h_{jr} + b^{(u)}\right)\\
c_j &= i_j \odot u_j + f_{jl} \odot c_{jl} + f_{jr} \odot c_{jr}\\
h_j &= o_j \odot \tanh\left(c_j\right)
\end{aligned}
\tag{7.26}
$$

For the leaf nodes, the h variable is set to zero, and for the internal nodes, the x variable is set to zero. The outputs i_j, f_{jk}, and o_j represent the input, forget, and output gates, respectively. k is a binary variable which indicates the right or left child of the current node.

- **Answer representation:** Each answer is represented by four answer aspects including answer entity (a_e), answer relation (a_r), answer type (a_t), and answer context (a_c). There is a set of contexts for each answer, and the final context representation (e_c) for the answer is generated by averaging all of the contexts (e_{c_i}) in this set. Latent representation of the answer entity (e_e), answer relation (e_r), answer type (e_t), and answer context (e_c) is generated. If the main entity is connected to the answer entity directly with one relation, the relation representation is considered as e_r; otherwise, e_r is the average of the representation of the relations in the path. The context set is formed by collecting the entities in the path between the main entity and the answer entity and their neighbors. Consider the example question "Where does PICC locate?" and the answer triple (*PICC, locate, Beijing*). The answer entity for *Beijing* is "e.112," the answer relation is *locate*, and the answer type is "city" which is extracted from triple (*Beijing, entity_type, city*).

  The TransE graph embedding method is used for generating the embedding of entities within the KG.

  A set of meta-path schemes are selected according to prior knowledge or expert advice. The meta-path schemes are represented as ($e_1 \rightarrow e_2 \rightarrow \cdots \rightarrow e_n$). A set of sequences are extracted from KG based on the meta-path schemes and are given to a Skip-gram model. They are also used in the embedding field.
- **Cross-attention:** Two different types of cross-attention, namely, answer-towards-question and question-towards-answer, are used for improving the representation of the question. Given each answer type, we need to re-read the question sentence to find the part of the question which represents the answer type. The answer-towards-question attention is based on this approach. On the contrary, when we attach the question parts together, we need to re-read the answer sentence several times in order to figure out the important parts.

This approach is covered by the question-towards-answer attention. These two attention mechanisms are explained in the following.

- **answer-towards-question attention:** The attention weights are calculated between each answer aspect and each node of the question's representation tree as follows:

$$\alpha_{ij} = \frac{\exp\left(w_{ij}\right)}{\sum_{k=1}^{n} \exp\left(w_{ik}\right)} \tag{7.27}$$

$$w_{ij} = \tanh\left(W^T\left[h_j; e_i\right] + b\right) \tag{7.28}$$

where α_{ij} represents the attention weight between the representation of tree node h_j and the aspect of answer a_i. h_j can be one of the $\left\{h_{\text{left}}, h_{\text{root}}, h_{\text{right}}\right\}$. A representation of the question is generated for each aspect of the answer. For example, the representation of the question based on the type aspect is calculated as follows:

$$v_q^t = \alpha_1^{\text{type}} h_{left} + \alpha_2^{\text{type}} h_{\text{root}} + \alpha_3^{\text{type}} h_{\text{right}} \tag{7.29}$$

The other representations of the question based on the other aspects (v_q^r, v_q^e, v_q^c) are calculated same as the above function. After generating the question's representation based on different aspects, the similarity of them with the corresponding aspect is calculated as $S\left(v_q^e, e_e\right), S\left(v_q^r, e_r\right), S\left(v_q^t, e_t\right)$, and $S\left(v_q^c, e_c\right)$. The matching score between the question and the candidate answer can be calculated by summing the above similarity scores.

- **question-towards-answer attention:** The question sentence should attend to different aspects of the answer sentence with different weights. Based on this, the question-toward-attention weights for each aspect of the answer ($\beta_{e_e}, \beta_{e_r}, \beta_{e_t}, \beta_{e_c}$) are calculated as follows:

$$\begin{aligned}
\beta_{e_i} &= \frac{\exp\left(w_{e_i}\right)}{\sum_{e_k \in e_e, e_r, e_t, e_c} \exp\left(w_{e_k}\right)} \\
w_{e_i} &= \tanh\left(W^T\left[\bar{q}; e_i\right] + b\right) \\
\bar{q} &= \frac{1}{3}\left(h_{left} + h_{\text{root}} + h_{right}\right)
\end{aligned} \tag{7.30}$$

Finally, the similarity score between the question and the answer is calculated as follows:

$$S(q, a) = \beta_{e_e} S\left(v_q^e, e_e\right) + \beta_{e_r} S\left(v_q^r, e_r\right) + \beta_{e_t} S\left(v_q^t, e_t\right) + \beta_{e_c} S\left(v_q^c, e_c\right) \tag{7.31}$$

- **Question-answer match:** For each pair of question and answer, m negative samples are collected. The negative samples are selected from the nearest entities to the answer entity. The negative samples are selected from the entities within the k-hop distance of the answer entity. k is the smallest number whose number of neighbor entities within the k-hop distance is more than m.

  The procedure of offline training is completed in the pair-wise form with the following loss function:

$$\mathcal{L}_{q,a,a'} = \sum_{a \in P_q} \sum_{a' \in N_q} \left[\gamma + S\left(q, a'\right) - S(q, a)\right]_+ \tag{7.32}$$

  where a and a' are correct and incorrect answers, respectively; γ is margin; and $[x]_+$ returns x if x is a positive number; otherwise, it returns zero. The loss function used for minimization is below.

$$\min \sum_q \frac{1}{|P_q|} \mathcal{L}_{q,a,a'} \tag{7.33}$$

- **Online response:** For finding the answer of the given question, the similarity score is calculated for each candidate answer within the candidate answer set, and the answer with the highest similarity score is returned. This approach is not applicable for the questions with more than one correct answer. To mitigate the abovementioned problem, the following answers are also extracted to form the final answer set $\hat{A}_q$.

$$\hat{A}_q = \left\{\hat{a} \mid S_{\max} - S(q, \hat{a}) < m\right\} \tag{7.34}$$

  where $\hat{A}_q$ is a candidate answer and m is margin.

  As the size of candidate answers is very large for each question, some methods have also been employed to reduce the size of candidate answer set.

The **PullNet** (Sun et al., 2019) model is designed to answer the non-trivial questions from two KB and text resources. PullNet follows an iterative process to generate a subgraph which is specifically generated for answering the question. The initial subgraph is created using the information from the question. In each iteration, the nodes from the current subgraph are selected for expansion based on the text and KG resources. We will explain the PullNet in more detail in the following.

- **The question subgraph**: The subgraph of question q is represented as $\mathcal{G}_q = \{\mathcal{V}, \mathcal{E}\}$ where $\mathcal{G}_q$ is constructed using the KB and text corpus. $\mathcal{V}$ represents a set of three different vertices (or nodes) including fact nodes $\mathcal{V}_f$, text nodes $\mathcal{V}_d$, and entity nodes $\mathcal{V}_e$. The entities of the KG are represented by entity nodes. The documents which are extracted from the text corpus are denoted by text nodes. Each document is represented as a sequence of words $w_1, ..., w_{|d|}$. The entity mentions within the documents are extracted using an entity linker. The facts

Algorithm 6 PullNet (Sun et al., 2019)

Initialize question graph G_q^0 with question q and question entities, with $\mathcal{V}^0 = \{e_{q_i}\}$ and $\mathcal{E}^0 = \emptyset$.

for $t = 1$ to T **do**
  Classify and select the entity nodes in the graph with probability larger than ϵ
  $\{v_{e_i}\} = classify\_pullnodes(G_q^t, k)$
  for all v_e in $\{v_{e_i}\}$ **do**
    Perform pull operation on selected entity nodes
    $\{v_{d_i}\} = pull\_docs(v_e, q)$
    $\{v_{f_i}\} = pull\_facts(v_e, q)$
    for all v_d in $\{v_{d_i}\}$ **do**
      Extracted entities in new document nodes
      $\{v_{e(d)_i}\} = pull\_entities(v_d)$
    end for
    for all v_f in $\{v_{f_i}\}$ **do**
      Extract head and tail of new fact nodes
      $\{v_{e(f)_i}\} = pull\_headtail(v_f)$
    end for
  end for
  Add new nodes and edges to question graph
  $G_q^{t+1} = update(G_q^t)$
end for
Select entity node in final graph that is the best answer
$v_{\text{ans}} = classify\_answer(G_q^T)$

from KG (v_s, r, v_o) are represented by fact nodes. The edges are indicated by $\mathcal{E}$. The edges connect the fact nodes and the text nodes to the entity nodes. A fact node v_f is connected to an entity node v_e if the entity node occurs as the subject or object of the fact. An entity node v_e is connected to a text node v_d if the entity occurs in the text.

- **Iterative subgraph expansion and classification**: The algorithm used for expanding the question subgraph and classification is represented in Algorithm 6. The question subgraph is initialized by the entities mentioned in the question and an empty set of nodes $\mathcal{G}_q^0 = \{\mathcal{V}^0, \mathcal{E}^0\}$ and $\mathcal{V}^0 = \{e_{qi}\}$. In each iteration, k entities are selected for expansion. To expand each entity, a set of relevant documents and facts are extracted. The entities of the relevant documents are extracted using an entity linking model. The head and tail entities of the relevant facts are also extracted. Then, the question's subgraph is updated by adding the new entities. Finally, a classification model is used to predict the answer entity.

  – **Pull operations**: The pull operations extract information from texts or KG by extracting entities from KG facts or documents. Given the document d as input, the $pull\_entities(v_d)$ function applies an entity linking model and returns all of the entities mentioned in the document d. Also, given the fact node v_f, the $pull\_headtail(v_f)$ function returns the subject and object entities of the input fact. The relevant documents are extracted from corpus by $pull\_docs(v_e, q)$ function using an IDF-based retrieval model (Michael et al., 2010). At first, the entity linker model is applied to the sentences of documents. The similarity of the documents to the input question is calculated, and the top N_d documents which have a mention to entity v_e are extracted. To retrieve the relevant facts, the facts with the v_e entity as their subject or object are selected as the candidates. Then, the similarity of the input question q and the fact's relation r is calculated to select the most similar fact. Similarity of the question q and a fact is calculated by dot product of the question's representations h_q and fact's embedding h_r. Representation of the question is generated using an LSTM. The last hidden of the LSTM is considered as the question's representation. A sigmoid function is applied to the result of dot product to convert it into a range of [0, 1] as follows:

$$
\begin{aligned}
h_q &= \text{LSTM}\left(w_1, \ldots, w_{|q|}\right) \in \mathbb{R}^n \\
S(r, q) &= \text{sigmoid}\left(h_r^T h_q\right)
\end{aligned} \tag{7.35}
$$

  - **Classify operations**: The graph CNN mode is applied to generate the representation of the nodes. The classification models are only applied on the entity nodes for two purposes: (1) selecting the nodes for expansion ($classify\_pullnodes$) and (2) classifying the nodes as the answer ($classify\_answer$). To construct the subgraph, in each step, k nodes are selected to expand using the $classify\_pullnodes(G_q^t)$ method. After the construction of the graph is completed, $classify\_answer(G_q^t)$ method is applied to calculate the probability of being a correct answer for each node. The same CNN architecture which is proposed by GRAFT-Net (Sun et al., 2018) is utilized for classifying. GRAFT-Net is capable of handling heterogeneous graphs which makes it suitable for this task.
  - **The update operation**: The subgraph of the previous step is passed to this operation, and it updates the graph by adding the recently extracted nodes. The extracted nodes can be text nodes, fact nodes, and entity nodes which are represented as $\{v_{d_i}\}$, $\{v_{f_i}\}$, and $\{v_{e(f)_i}\} \cup \{v_{e(d)_i}\}$, respectively. The set of edges $\mathcal{E}$ is also updated by the (1) edges of newly retrieved text and fact nodes and (2) the edges which connect the previous nodes to the new nodes.

- **Training**: The training data includes the questions and the answer entities which is not sufficient for training the middle models like classifiers and similarity measurement models. To mitigate this problem, an approach based on weak supervision is utilized. For given question and answer entities, an estimation of the question graph is constructed. This graph is constructed by connecting the question entities to the answer entities via the shortest path. When training the $classify\_pullnodes$ operation in step t, the nodes in the distance of $t + 1$ of the intermediate nodes are considered as the grand truth for expansion. In addition, the positive relations which connect the current nodes to the candidate nodes in distance $t + 1$ are used to train the similarity score $S(h_r, q)$. Teacher forcing is applied for training the model. In the process of training and pulling, the nodes

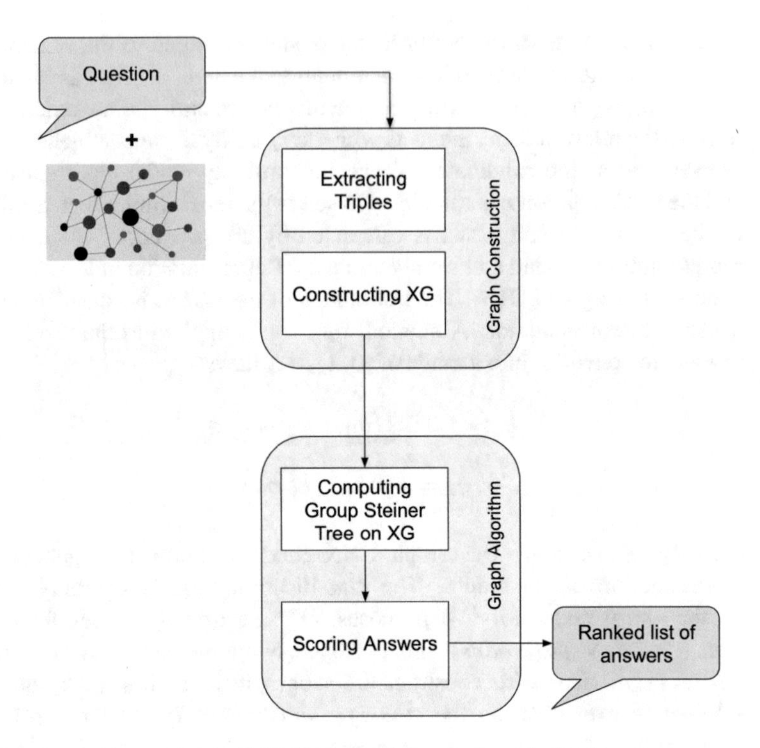

Fig. 7.6 The architecture of the UniQORN model (Pramanik et al., 2021)

with a score higher than a pre-defined threshold ϵ are also selected besides the top k nodes.

The architecture of the **UniQORN** model (Pramanik et al., 2021) is shown in Fig. 7.6. This model combines the textual information and KG information in the first step to build a graph. The model then selects the answer from the graph by using the group Steiner tree (GST) algorithm. A detailed explanation of this model is presented in the following.

- **Building the context graph (XG):** The context graph is created from two knowledge sources including the KG and text corpus. The way of constructing each element of XG using two knowledge sources is shown in Table 7.2. We use the question "Director of the western for which Leo won an Oscar?" as an example to explain the curation of XG from text and KG in the following.

  1. XG from knowledge graph: To form the XG from KG, the entities of the question sentence ($E(q)$) are extracted using Named Entity Recognition (NER) and Named Entity Disambiguation (NED). For the example question, the entities of the KG are as (Leonardo DiCaprio, Academy Awards). As for the short inputs, the NED is vulnerable to error; the NER is also used to identify the entities. A dictionary with entities as key and names as value

Table 7.2 Producing different factors of XGs from KG and text corpus

Scenario	KG	Text
Triples in XG	NER + KG-lookups	Retrieval system + Open IE
Node types	Entity, predicate, class, literal	Entity, predicate, class
Class nodes	instance of triples	Hearst patterns
Edge types	Triple, alignment, class	Triple, alignment, class
Entity alignments	With KG aliases	With entity-mention lexicons
Predicate alignments	With word embeddings	With word embeddings
Node weights	W.r.t. question tokens	W.r.t. question tokens
Edge weights	Degenerate	With term proximity and lexicons

is utilized to get the top 5 possible entities (Leo, Leonardo DiCaprio, pope Leo I). $E(q)$ is generated using these entities. In the next step, all of the facts in the one-hop distance of the extracted entities are selected to form the $E(q)_j$. Finally, the context graph $XG^K(q)$ is generated by selecting the largest connected component (LCC) of the graph. By selecting the LCC, the facts related to the context of the question are preserved, and non-related facts (like facts about pope Leo I) are discarded.

2. XG from text corpus: A set of related documents to the input question (q) are gathered to form a document set D. The documents can be collected from search engines or IR systems. The structure of the documents is converted to the structure of KG by using Open IE (Mausam, 2016). This helps us to process the documents like the KGs. By applying the Open IE model, we have a set of SPO triples with brief text phrases as its components. According to the limitation of the available models, a custom Open IE is developed. At first, NER, POS tagging, and coreference resolution are applied to the input sentence. Some patterns for entities and predicates are defined as a sequence of POS and are applied to each sentence $S_j \in D$ of the documents to produce $S_j \equiv E_1 \dots E_2 \dots P_1 \dots E_3 \dots P_2 \dots E_4$. Then, the entities which are directly connected to each other by one predicate are selected (E_{i_1}, E_{i_2}). In this step, we aim at achieving higher recall because the noise will be eliminated in the further steps.

   The Hearst patterns model (Hearst, 1992) is applied to extract the entity class information. This model works by patterns like "NP_2 such as NP_1" and "NP_1 is an NP_2" (e.g., "western films such as The Revenant" and "The Revenant is a 2015 American western film"). The output of the model is in the form of $< NP_1$, class, $NP_2 >$ and is added to the triple collection. Finally, all of the triples are connected to each other, and LCC is applied to form the XG of text $XG^D(q)$.

- **Answering over the context graph:** At first, the anchor nodes are recognized for retrieving the candidate answers $a \in A$ of the context graph $XG(q)$. Weight of each node within the graph is calculated based on the similarity of the node labels to the question tokens. The anchor nodes $\mathcal{A}_k$ are the nodes with a weight higher than the specified threshold τ_{anchor}. The anchor nodes are classified into classes $\{\mathcal{A}_k\}$. The nodes in one class correspond to the same question token. After finding the anchor nodes, the context graph is represented as $XG^{(\cdot)}(q) = (N, \mathcal{E}, N^T, \mathcal{E}^T, N^W, \mathcal{E}^W)$ where N^T, $\mathcal{E}^T$, N^W, and $\mathcal{E}^W$ represent the node type, entity type, node weight, and entity weight, respectively. We explain the next steps in the following.

1. Group Steiner tree: The rules for identifying a good answer are defined as follows: (1) the path which connects the anchors includes the answers, (2) the probability of including the correct answer for a path is directly correlated to the weight of path and reversely correlated to the lengths of path, and (3) the path must include at least one anchor node from each group. To solve this problem, the general concept of group Steiner trees (GST) (Chanial et al., 2018; Li et al., 2016; Shi et al., 2020) is used. The GST problem is defined as finding a Steiner tree with minimum cost which includes at least one terminal (node) from each group, while the edge weight is defined as $C(\mathcal{E}_k) = 1 - W^{\mathcal{E}}(\mathcal{E}_k)$. Several approximation algorithms (Garg et al., 1998) are proposed for solving the GST problem as it is an NP-hard problem. The algorithm proposed by Ding et al. (2007) is adapted in this work. The dynamic programming approach is utilized in this algorithm.
2. Relaxation to top-k GSTs: The answers are in entities, but the GST can join a terminal from each anchor group while it has no internal nodes, or predicates or classes are internal nodes. To avoid the mentioned problem, we need to change our problem to the problem of finding the top-k least-cost GSTs. In this way, not only we will ensure having a non-empty answer list, but also we have a natural way to rank the answers. For example, if an answer exists in several low-cost GSTs, its score can be increased.
3. Answer ranking: The non-terminal nodes are candidate answers. The candidate answers are ranked according to the number of GSTs which include them. An additional post-processing step is performed because the entities are not canonicalized. In the post-processing step, answers are merged if (1) they are subsequences of another answer or (2) they have a common alignment edge.

We have reviewed several KBQA models which use textual data for enhancing the performance. These models use the textual information for various objectives. The Text2KB model uses the textual data to provide context information for a short question. The context information is used for entity linking and identifying the entities. In addition, the textual information is used for estimating the association of entities, enriching the representation of entities, and ranking the candidate answers.

The hybrid QA model uses both textual data and KG for capturing the meaning of the question, entity linking, and relation extraction. In addition, the textual data is used for solving the limitations of KBQA that are due to the closed vocabulary of predicates.

The GRAFT-Net model builds a question-specific subgraph using textual data and information from KG. Also, it uses GRAFT-Net for generating the node representation and calculating the probability of being a correct answer for each node.

The KG-QA model uses textual data in generating word embeddings and constructing the context of the answer.

The PullNet model constructs a subgraph for the given question in an iterative process. The subgraph is constructed using the data from text and KG.

The UniQORN model also utilizes both data sources in generating a subgraph for answering the input question. It uses the group Steiner tree (GST) model for selecting the answer from subgraph.

As can be seen, the available KBQA models can use the textual data for (1) creating the question's subgraph, (2) obtaining the context information for the input question, (3) entity linking and identification, and (4) generating embeddings.

7.3 Summary

We devoted this chapter to review the KBQA which benefits from textual data for improving the performance of KBQA. We have presented a detailed explanation of the architecture of the models while emphasizing on the way they incorporate the textual data in their architecture.

References

Angeli, G., Premkumar, M. J. J., & Manning, C. D. (2015). Leveraging linguistic structure for open domain information extraction. In *Proceedings of the 53rd Annual Meeting of the Association for Computational Linguistics and the 7th International Joint Conference on Natural Language Processing (Volume 1: Long Papers)*, Beijing, China (pp. 344–354). Association for Computational Linguistics. https://doi.org/10.3115/v1/P15-1034

Chanial, C., Dziri, R., Galhardas, H., Leblay, J., Le Nguyen, M. H., & Manolescu, I. (2018). Connectionlens: Finding connections across heterogeneous data sources. *Proceedings of the VLDB Endowment, 11*(12), 2030–2033. https://doi.org/10.14778/3229863.3236252.

Ding, B., Yu, J. X., Wang, S., Qin, L., Zhang, X., & Lin, X. (2007). Finding top-k min-cost connected trees in databases. In *2007 IEEE 23rd International Conference on Data Engineering* (pp. 836–845). IEEE.

Fader, A., Zettlemoyer, L., & Etzioni, O. (2013). Paraphrase-driven learning for open question answering. In *Proceedings of the 51st Annual Meeting of the Association for Computational Linguistics (Volume 1: Long Papers)*, Sofia, Bulgaria (pp. 1608–1618). Association for Computational Linguistics. https://www.aclweb.org/anthology/P13-1158.

Garg, N., Konjevod, G., & Ravi, R. (1998). A polylogarithmic approximation algorithm for the group Steiner tree problem. In *Proceedings of the Ninth Annual ACM-SIAM Symposium on Discrete Algorithms, SODA '98*, USA (pp. 253–259). Society for Industrial and Applied Mathematics. ISBN 0898714109.

Haveliwala, T. H. (2002). Topic-sensitive pagerank. In *Proceedings of the 11th International Conference on World Wide Web, WWW '02*, New York, NY, USA (pp. 517–526). Association for Computing Machinery. ISBN 1581134495. https://doi.org/10.1145/511446.511513

Hearst, M. A. (1992). Automatic acquisition of hyponyms from large text corpora. In *COLING 1992 Volume 2: The 14th International Conference on Computational Linguistics*. https://aclanthology.org/C92-2082

Li, R.-H., Qin, L., Yu, J. X., & Mao, R. (2016). Efficient and progressive group Steiner tree search. In *Proceedings of the 2016 International Conference on Management of Data, SIGMOD '16*, New York, NY, USA (pp. 91–106). Association for Computing Machinery. ISBN 9781450335317. https://doi.org/10.1145/2882903.2915217

Liang, C., Berant, J., Le, Q., Forbus, K. D., & Lao, N. (2017). Neural symbolic machines: Learning semantic parsers on freebase with weak supervision. In *Proceedings of the 55th Annual Meeting of the Association for Computational Linguistics (Volume 1: Long Papers)*, Vancouver, Canada (pp. 23–33). Association for Computational Linguistics. https://doi.org/10.18653/v1/P17-1003

Mausam, M. (2016). Open information extraction systems and downstream applications. In *Proceedings of the Twenty-Fifth International Joint Conference on Artificial Intelligence, IJCAI'16* (pp. 4074–4077). AAAI Press. ISBN 9781577357704

Michael, M., Erik, H., & Otis, G. (2010). *Lucene in action: Covers apache lucene 3.0*. Manning Publications.

Pramanik, S., Alabi, J., Roy, R. S., & Weikum, G. (2021). Uniqorn: Unified question answering over RDF knowledge graphs and natural language text. arXiv preprint arXiv:2108.08614.

Savenkov, D., & Agichtein, E. (2016). When a knowledge base is not enough: Question answering over knowledge bases with external text data. In *Proceedings of the 39th International ACM SIGIR Conference on Research and Development in Information Retrieval*, SIGIR '16, New York, NY, USA (pp. 235–244). Association for Computing Machinery. ISBN 9781450340694. https://doi.org/10.1145/2911451.2911536

Savenkov, D., Lu, W. L., Dalton, J., & Agichtein, E. (2015). Relation extraction from community generated question-answer pairs. In *Proceedings of the 2015 Conference of the North American Chapter of the Association for Computational Linguistics: Student Research Workshop*, Denver, Colorado (pp. 96–102). Association for Computational Linguistics. https://doi.org/10.3115/v1/N15-2013

Schlichtkrull, M., Kipf, T. N., Bloem, P., Berg, R. V. D., Titov, I., & Welling, M. (2018). Modeling relational data with graph convolutional networks. In *European Semantic Web Conference* (pp. 593–607). Springer.

Shi, Y., Cheng, G., & Kharlamov, E. (2020). *Keyword search over knowledge graphs via static and dynamic hub labelings* (pp. 235–245). Association for Computing Machinery. https://doi.org/10.1145/3366423.3380110.

Socher, R., Huang, E., Pennin, J., Manning, C. D., & Ng, A. (2011). Dynamic pooling and unfolding recursive autoencoders for paraphrase detection. In *Proceedings of the 24th International Conference on Neural Information Processing Systems, NIPS'11* (pp. 801–809). Curran Associates. ISBN 978-1-61839-599-3

Sun, H., Dhingra, B., Zaheer, M., Mazaitis, K., Salakhutdinov, R., & Cohen, W. W. (2018). Open domain question answering using early fusion of knowledge bases and text. In *Proceedings of the 2018 Conference on Empirical Methods in Natural Language Processing*, Brussels, Belgium (pp. 4231–4242). Association for Computational Linguistics. https://doi.org/10.18653/v1/D18-1455

Sun, H., Bedrax-Weiss, T., & Cohen, W. W. (2019). Pullnet: Open domain question answering with iterative retrieval on knowledge bases and text. In *Proceedings of the 2019 Conference on Empirical Methods in Natural Language Processing and the 9th International Joint Conference on Natural Language Processing (EMNLP-IJCNLP)* (pp. 2380–2390).

Tong, P., Zhang, Q., & Yao, J. (2019). Leveraging domain context for question answering over knowledge graph. *Data Science and Engineering, 4*(4), 323–335.

Xu, K., Feng, Y., Huang, S., & Zhao, D. (2016a). Hybrid question answering over knowledge base and free text. In *Proceedings of COLING 2016, the 26th International Conference on Computational Linguistics: Technical Papers*, Osaka, Japan (pp. 2397–2407). The COLING 2016 Organizing Committee. https://www.aclweb.org/anthology/C16-1226

Xu, K., Reddy, S., Feng, Y., Huang, S., & Zhao, D. (2016b). Question answering on Freebase via relation extraction and textual evidence. In *Proceedings of the 54th Annual Meeting of the Association for Computational Linguistics (Volume 1: Long Papers)*, Berlin, Germany (pp. 2326–2336). Association for Computational Linguistics. https://doi.org/10.18653/v1/P16-1220

Yang, Y., & Chang, M. W. (2015). S-MART: Novel tree-based structured learning algorithms applied to tweet entity linking. In *Proceedings of the 53rd Annual Meeting of the Association for Computational Linguistics and the 7th International Joint Conference on Natural Language Processing (Volume 1: Long Papers)*, Beijing, China (pp. 504–513). Association for Computational Linguistics. https://doi.org/10.3115/v1/P15-1049

Zhang, H., Lu, G., Zhan, M., & Zhang, B. (2022). Semi-supervised classification of graph convolutional networks with Laplacian rank constraints. *Neural Processing Letters, 54*(4), 2645–2656. https://doi.org/10.1007/s11063-020-10404-7

Chapter 8
Question Answering in Real Applications

Abstract We believe that the future direction and trends in QA are toward real systems. In this chapter, some real applications of QA will be introduced. We will discuss the architecture of these full pipeline QA approaches as well as their applications. Considering the advantages of both textual data and knowledge bases, the real applications of QA aim to benefit from both sources. Therefore, in this chapter, we will see how a combination of both approaches can be used in real scenarios.

8.1 IBM Watson

Watson (Ferrucci et al., 2010) is a QA system with great precision, confidence, and speed at the level of a human expert. Watson is the outcome of IBM researchers' efforts to build a computer system that can compete on the TV quiz show, *Jeopardy!*, in real time with humans. To this aim, they developed the DeepQA architecture and implemented Watson based on this architecture. In the following, we will introduce the TV quiz show, *Jeopardy!*, DeepQA architecture, and baselines used for the evaluation of Watson.

8.1.1 The Jeopardy! *Challenge*

In *Jeopardy*!, three participants compete by answering the questions asked in natural language. The competition includes three rounds. The questions, asked in *Jeopardy!*, are from various domains and subjects, and in the case of answering the questions wrongly, the competitor receives a penalty.

A board, representing 30 clues from 6 different categories, is shown to competitors. The clues of each column are sorted in ascending order based on their value in dollars. This board is used in the first two rounds, and the winner is asked to select a category and value. After selecting the value and category, the presenter reads the corresponding clue to competitors. When the presenter finishes reading the clue, the

S. Momtazi, Z. Abbasiantaeb, *Question Answering over Text and Knowledge Base*,
https://doi.org/10.1007/978-3-031-16552-8_8

competitors are allowed to buzz in for answering the clue. The presenter reads the question, which usually takes 1 to 6 seconds, and during this time, the competitors must think to find the answer and determine whether they are confident about the answer or not. The competitor who first buzzes in must answer the question. In the case of giving the correct answer, the value of the answered clue will be added to the score of the competitor, and he/she has the permission to select the next clue from the board. Otherwise, the score of the competitor will be diminished. The answer to clues must be in the form of a question. An example of a clue with the correct response (answer) is as follows:

Clue: "This drug has been shown to relieve the symptoms of ADD with relatively few side effects."
Answer: "What is Ritalin?"

To avoid getting a penalty, the competitors must answer the questions with high confidence; hence, accurate computation of the confidence is highly important. The competitors must be smart, have good knowledge, and be quick. One special hidden clue named Daily Doubles exists in the board that will be unfolded after selecting the clue. The Daily Doubles clue must be answered by the competitor who selected the clue, and its value is determined by himself/herself. If the competitor gives the wrong answer, the specified value will be diminished from his/her score.

The last round is pretty different, and the competitors gamble on questions. At first, the competitors gamble on questions. At first, the competitors select the category, and then, before the clue appears, they specify their bet privately. They have 30 seconds to prepare their answers.

The subject of the categories ranges from topics that bring more information about scope of the answer including history and politics to other topics that cannot help to limit the available resources. Category of the clues can be necessary for answering the question or not. Sometimes, not only the categories are not useful, but also they can be misleading. Determining the purpose of the clue and the effective parts of the clue for figuring out the intended answer to the question is a challenging task. Some clues are more complex and require more reasoning to answer. They include multiple facts about the answer that are essential for answering the clue. An example of a complex clue is as follows:

Category: "Rap" Sheet
Clue: This archaic term for a mischievous or annoying child can also mean a rogue or scamp.
Subclue 1: This archaic term for a mischievous or annoying child.
Subclue 2: This term can also mean a rogue or scamp.
Answer: Rapscallion

To answer the above clue, it is less probable to find a sentence which includes both the first and the second subclues. Consequently, we search for each subclue independently and select the common answer in answer of both subclues as the answer of the question.

In another type of complex clues, the answer of a subclue is required in constructing the other subclue. In other words, the outer subclue contains the inner subclue and needs its answer to be completed. For example, consider the following clue:

Category: Diplomatic Relations
Clue: Of the four countries in the world that the United States does not have diplomatic relations with, the one that's farthest north.
Inner subclue: The four countries in the world that the United States does not have diplomatic relations with (Bhutan, Cuba, Iran, North Korea).
Outer subclue: Of Bhutan, Cuba, Iran, and North Korea, the one that's farthest north.
Answer: North Korea

It does not seem reasonable to search for a sentence which includes the exact answer of the above clue. The above clue is broken into two inner and outer subclues. The outer subclue cannot be answered independently and needs the answer of the inner clue. Therefore, the first subclue is replaced with its answer to complete the outer subclue. The two mentioned types of clues are decomposable clues. Another type of the clues which must be answered by decomposing is the puzzles category. The before and after category is one famous example of puzzle categories in which subclues have a common word. An example of the before and after category is as follows:

Category: Before and After Goes to the Movies
Clue: Film of a typical day in the life of the Beatles, which includes running from bloodthirsty zombie fans in a Romero classic.
Subclue 2: Film of a typical day in the life of the Beatles.
Answer 1: (A Hard Day's Night)
Subclue 2: Running from bloodthirsty zombie fans in a Romero classic.
Answer 2: (Night of the Living Dead)
Answer: A Hard Day's Night of the Living Dead

Based on the agreement of IBM and *Jeopardy!* Productions, Inc., two kinds of questions including audiovisual (A/V) and Special Instructions questions are eliminated. In the audiovisual questions, the clue contains part of a video or audio. An example of a Special Instructions question is as follows:

Category: Decode the Postal Codes
Verbal instruction from host: We're going to give you a word comprising two postal abbreviations; you have to identify the states.
Clue: Vain
Answer: Virginia and Indiana

The distribution of the clues' subject in the *Jeopardy!* challenge is determined using 20,000 instances of questions by extracting the Lexical Answer Type (LAT) of the questions. The LAT is a term within the clue sentence which specifies the type of

Table 8.1 Most common types among top 300 *Jeopardy!* answers

Answer type	Frequency
Country	8000–10000
State	4000–6000
City	2000–4000
Historical figure	2000–4000
President	2000–4000
Author	0–2000
Planet	0–2000
Number	0–2000
Fictional character	0–2000
Element/molecule	0–2000

the answer. About 12% of the clues do not contain the exact LAT word, and it must be inferred from context. The scatter plot of LATs is depicted in Table 8.1.

Winning in the *Jeopardy!* challenge relies on multiple factors including precision of QA component, speed, betting strategy, confidence measurement, and criteria of selecting clues. The correctness, speed, and confidence of Watson are calculated by precision and percent metrics. The precision is defined as the ability of the system to answer the questions correctly, and the percent is defined as a portion of the questions that the system decides to answer. The system calculates a confidence value to decide whether to answer the clue or not. When the confidence score is very high, the system answers fewer questions with high precision. By decreasing the confidence threshold, the system answers more questions with lower precision.

8.1.2 DeepQA

DeepQA has a probabilistic parallel architecture with accompanying methodology. DeepQA is not limited to *Jeopardy!* challenge and is used in other QA tasks like TREC-QA. Different techniques are adapted to prepare DeepQA for the *Jeopardy!* challenge. The subtle point besides choosing the best technique is combining them in a way that their overall performance will improve the metrics including confidence, accuracy, and speed. The main principles of constructing DeepQA are as follows:

- Massive parallelism: Utilized for processing several hypotheses and interpretations.
- Many experts: Helps to integrate, apply, and evaluate various probabilistic questions.
- Pervasive confidence estimation: The confidence scores estimated by each component are combined for measuring the final confidence score.
- Integrate shallow and deep knowledge

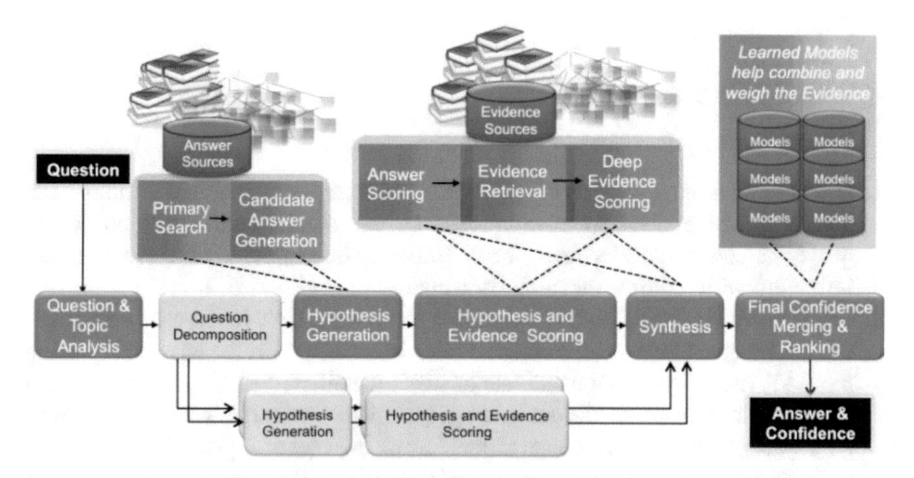

Fig. 8.1 The architecture of the DeepQA framework which is used in IBM Watson ("©Creative Commons Attribution license, reprinted with permission")

A high-level architecture of DeepQA is shown in Fig. 8.1. This architecture is explained in the following sections.

8.1.2.1 Content Acquisition

In this step, the required content or documents for answering the input question are gathered both manually and automatically. The types of the example questions and features of their application domain are extracted to describe questions that must be answered. The example questions are analyzed manually, and the domain analysis is performed by statistical models like the LAT analysis. The contents are extracted from articles, dictionaries, encyclopedias, thesauri, and literary works.

The DeepQA uses an offline corpus which is expanded by answering a question. Besides the unstructured data sources, other semi-structured and structured data sources including YAGO, WordNet, and DBpedia are identified and gathered. This makes the system a hybrid model which uses both textual and knowledge base sources.

8.1.2.2 Question Analysis

The question sentence is processed to determine what the question sentence expects in the answer. In Watson, the following processes are performed:

- Semantic role labeling,
- Coreferences resolution,
- Shallow and deep parsing,

- Logical forms extraction,
- NER,
- Relations detection.

The above analysis is used in the next components.

1. Question classification: The type of the question and some segments of the question, which need specific processing, are determined. The type of the questions can be puzzle, math, or definition. The segments of a question can be a word with multiple meanings or a clue.
2. Focus and LAT detection: As explained previously, LAT is a word or a phrase within the question sentence which determines the type of the answer. By using the LAT, many irrelevant candidate answers are filtered easily with no need to model their semantics. Various answer type detection models are examined in the DeepQA project, while each model has its own set of types. The best way to utilize the pre-defined answer type detection models is to create a mapping from their answer types to the existing answer types.

   Focus of the question is part of the question that, when is substituted by the answer, generates a complete and meaningful sentence. Focus of the question carries important information about the answer and usually is the subject or object of the clue. For example, in the clue "This title character was the crusty and tough city editor of the Los Angeles Tribune," the phrase "This title character" is the focus of the question.
3. Relation detection: A large number of questions include semantic or subject-relation-object relations. Watson utilizes relation detection during the QA process to directly extract the answer from knowledge graphs. The domain of the questions asked in *Jeopardy!* covers a variety of relations that are stated in various ways. As a result, it is not easy to create and manage a database for querying the facts of *Jeopardy!* clues. Utilizing existing knowledge graphs depends on two factors: (1) analyzing the question and (2) extracting the relations from knowledge graphs. Watson uses Freebase for retrieving the relations where in the case of having the highest recall, the most frequent relations are detected accurately and the less frequent relations remain a challenge. In a sample of 20,000 questions from *Jeopardy!*, 25% of the questions have frequent relations.
4. Decomposition: As explained in the previous section, some *Jeopardy!* questions are answered by decomposition. The questions are classified to realize whether they must be decomposed or not by statistical classification methods in DeepQA and the best decomposition of the input questions into several subquestions is selected by utilizing rule-based deep parsers.

   As presented in Fig. 8.1, the decomposed questions are solved by parallel computation, and the final answer of the parallel components are integrated through the synthesis process. The parallel computation is represented by light gray components. In the case of having nested clues, the process is called recursively. The end-to-end QA system is repeated recursively from the inner subclue to the outer subclue.

8.1.2.3 Hypothesis Generation

The output of question analysis is passed to the hypothesis generation component for retrieving the candidate answers. The candidate answers are called hypotheses, and they are retrieved from resources through the primary search. The snippets with the size of the answer are extracted from resources to form the hypothesis. A confidence score must be calculated for each hypothesis to indicate its correctness. In Watson, a combination of various techniques is used for primary search and generating the candidate answers.

1. Primary search: The goal of primary search is to retrieve all of the contents that possibly contain the correct answer. Consequently, achieving a higher recall is very important. In the following steps, the confidence score of the retrieved candidate answers is calculated. Various text search algorithms are used for primary search including document search and passage search. Also, SPARQL is used for searching the triples in knowledge bases. For searching the triples, the named entities of the question are used, and the triples with related entities to the question's entities are selected. The optimal primary search defined for DeepQA achieved 85% recall on the top 250 questions.
2. Candidate answer generation: Given the output of primary search, the candidate answers are generated according to the search method. For example, the candidate answers within the documents returned by document search from title-oriented resources can be generated using the title. The title itself can be considered as a candidate answer, and other candidate answers can be generated by substring and link analysis.

   To generate the candidates' answers from passages returned by passage search, some analysis including NER is needed. In the case of using dictionary lookup or knowledge bases, the candidate answers are produced directly from search results.

   The correct answer must be retrieved in this step as there is no further step to generate the candidate answers. As a result, higher recall value is preferred to precision, and the correctness (precision) of the retrieved candidate answers would be determined in the further steps.

8.1.2.4 Soft Filtering

The initial set of candidate answers is very large, and computing the intensive scoring functions over all of them is not efficient. To this aim, a lightweight scoring function is utilized for pruning the candidate answers by eliminating the candidate answers that are most likely wrong. The soft filtering score is computed for each candidate answer and the candidate answers which meet the threshold requirement of the soft filtering stage will be passed to the further scoring steps (hypothesis and evidence scoring) and the other ones will be forwarded to the last merging step.

8.1.2.5 Hypothesis and Evidence Scoring

A set of extra supporting evidences is collected for the candidate answers or hypothesis, and they are evaluated by various deep scoring models.

- Evidence retrieval: The extra supporting evidences are collected for better evaluation of candidate answers. The model uses a combination of various evidence retrieval models. An efficient method used for evidence retrieval is passage search. In the passage search, a query is generated based on the input question, and the candidate answer is added to the query as an essential term. This search helps to retrieve the passages related to the context of the input question that include the candidate answer. The supporting evidences gathered in this step are passed to the deep evidence scoring functions for evaluating the candidate answer according to the given supporting evidence.
- Scoring: The scoring function calculates the degree to which the collected evidence supports the candidate answer. In the DeepQA framework, various models or scorers are combined to consider different aspects of the gathered evidence for calculating the score. In Watson, about 50 scorer models are utilized for scoring the candidate answer based on the collected evidence. They generate scores in various formats including probabilities, counts, and categorical features. The evidences are collected from triples, structured text, and unstructured text. The scorers score the evidences and candidate answers by using various criteria including matching the structure of the passage and the question, reliability of passage's origin, taxonomic classification, and semantic and word-level relations.

  For example, consider the following clue with its answer and the evidence gathered in the previous step as follows:

  Clue: "He was presidentially pardoned on September 8, 1974"
  Answer: "Nixon"
  Evidence: "Ford pardoned Nixon on Sept. 8, 1974."

  Some of the scoring models used for scoring the above example are listed below.

  - Count of the common terms between the question and the retrieved passage,
  - Using sequence matching models for finding the longest subphrase among them,
  - Aligning the logical form of the passage and the question,
  - Geospatial and temporal reasoning,
  - Utilizing knowledge from triples.

  Finally, the scores generated by scorer models are combined to form an overall score.

8.1.2.6 Final Merging and Ranking

The goal of this step is extracting the best single hypothesis which answers the clue correctly and best matches the evidence, considering the output of various scoring methods.

8.1.2.7 Answer Merging

A clue can have several equal candidate answers with various surface forms. The answer candidates are merged before the ranking and confidence estimation step to prevent the ranker from comparing answer candidates with similar context and different surface forms. Watson utilizes a combination of matching, coreference resolution, and normalization algorithms to find the equal answer candidates. Then the set of equal answer candidates are merged by combining their scores per feature.

8.1.2.8 Ranking and Confidence Estimation

In this step, the confidence score of the merged hypothesis is calculated to rank them. A machine learning model can be trained for both calculating the confidence score and ranking the hypothesis given a QA dataset with its score. The task of estimating the score and ranking can be performed in two distinct phases. For the both tasks, the scores can be grouped based on their origin like type matching and passage scoring, and several intermediate models are used for training. A meta-learner is applied over the scores generated from intermediate models.

The meta-learner used in Watson includes specific models for handling the different types of questions like factoid and puzzle.

8.1.3 Baselines

The two baselines named Practical Intelligent QA Technology (PIQUANT) (Prager et al., 2004) developed at IBM and OpenEphyra (Schlaefer et al., 2006) developed at CMU are used for evaluating DeepQA in the *Jeopardy!* challenge. Both the baselines were designed for evaluating the TREC QA challenge. The TREC challenge is different from *Jeopardy!* in several ways including the following: (1) questions of the TREC are much simpler than the *Jeopardy!* clues, and (2) TREC had a small corpus as source for retrieving the answers, and it permitted utilizing the information on the Web, while in the *Jeopardy!* challenge, the system must contain the sources, and using the Web is not allowed.

The PIQUANT was modified to be prepared for only answering the clues in *Jeopardy!* challenge, while other aspects like betting and speed were ignored. Five hundred questions were collected from *Jeopardy!* clues, and a corpus containing the answer of more than 90% of the questions was given to PIQUANT for evaluation. The evaluation of the PIQUANT system revealed its 13% precision over all the clues and 47% precision over 5% of the most confident clues. The OpenEphyra achieved lower than 15% accuracy in evaluation.

8.2 DrQA

DrQA (Chen et al., 2017) is a QA system designed to detect the answer span in source documents. Despite the DeepQA, which utilizes diverse sources of information for extracting the answer, only the textual documents are used as information sources of the DrQA. The model must be very careful to find the answer from one information source because the answer may not exist in more than one document.

The overall architecture of DrQA, as is shown in Fig. 8.2, consists of two main components: (1) a document retriever that retrieves the related documents given the question sentence and (2) a document reader which extracts the answer span from retrieved documents. The document reader is a machine comprehension model which is termed as machine reading at scale.

The document retriever component uses a TF-IDF weighted bag of words model of the question sentence and documents to compare them. In another setting, the TF-IDF model is augmented by bigram counts. Given an input question, the document retriever component retrieves five relevant Wikipedia articles that with a high probability include the correct answer.

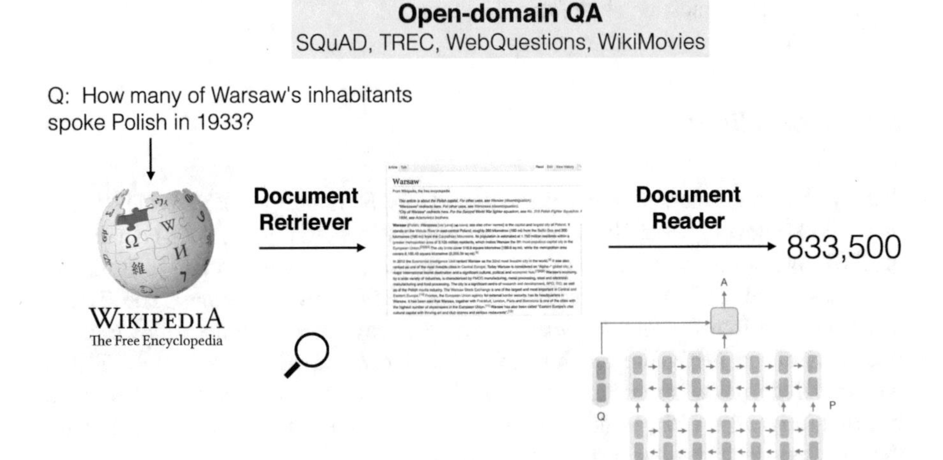

Fig. 8.2 DrQA QA system (Chen et al., 2017) ("©1963–2022 ACL, reprinted with permission")

In the document reader component, RNNs are used for modeling the semantic representation of the question and the paragraph. For encoding the paragraphs, several features including exact match, token feature, and aligned question embedding are used besides the word embedding.

After modeling the semantic representation of the question and paragraphs, two distinct prediction models are trained for predicting the start and end token of the answer's span.

Wikipedia is used as a collection of documents for retrieving the answers in DrQA, and it can be replaced with any other collection of documents. The structured sections of the Wikipedia pages are eliminated, and just the textual contents of the documents are preserved. Besides the Wikipedia articles, SQuAD QA dataset (Rajpurkar et al., 2016) is used for training the document reader. SQuAD is a reading comprehension dataset gathered from Wikipedia articles. Each paragraph has a set of questions. The answer of each question is shown in the paragraph by indicating its start and tokens.

CuratedTREC (Baudiš & Šedivỳ, 2015), WebQuestions (Berant et al., 2013), and WikiMovies (Miller et al., 2016) open-domain QA datasets are used for evaluating DrQA.

8.3 YodaQA

YodaQA (Baudiš & Šedivỳ, 2015) is an open-domain and modular pipeline for QA that relies on various paradigms of knowledge base for answering the questions. YodaQA returns an ordered list of answers for the input question. According to the architecture of YodaQA, which is shown in Fig. 8.3, it is mainly inspired by DeepQA framework and includes the following modules:

- **Question analysis:** In the question analysis component, several tasks including (1) recognizing named entities of the question, (2) POS tagging, and (3) forming the question's dependency parse tree are performed. Some QA features including Clues, Focus, and LAT are also extracted.

  Clues are important keywords of the question sentence that convey the meaning of the question and are used for querying the candidate answer. Each clue is given a weight, and when a clue is related to the name or alias of a Wikipedia page, its weight will be enhanced. Focus of the question represents the intent of the question, and it is extracted by using hand-crafted heuristics. LAT is an English word which explains the answer's type and is detected from focus terms.
- **Answer production:** The candidate answers are produced from both unstructured and structured answer sources. The main source of answer extraction for the unstructured data is *enwiki* corpus. Some articles are entitled with the name of an entity and describe it. Usually, the first line of these articles is a description of the corresponding entity. This property of *enwiki* corpus is used in answer

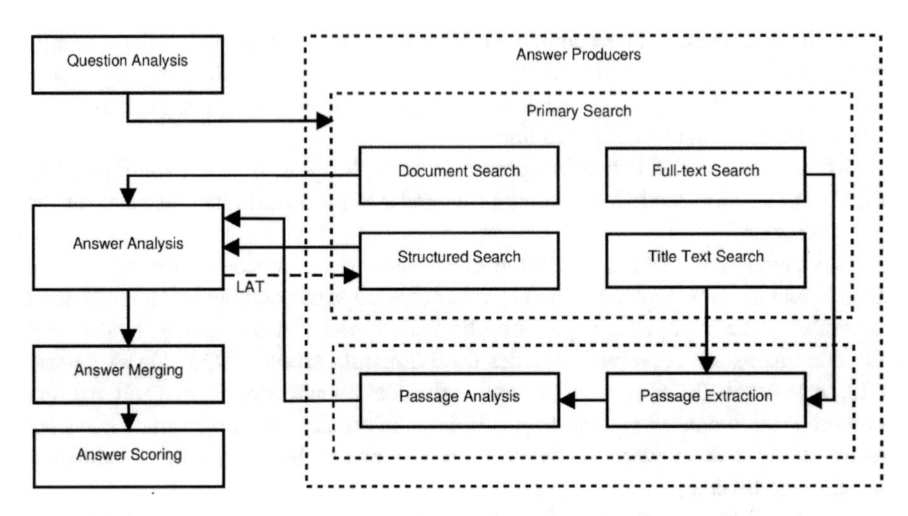

Fig. 8.3 The architecture of YodaQA pipeline for QA (Baudiš & Šedivý, 2015) ("©Springer International Publishing Switzerland 2015, reprinted with permission")

search. The search strategies used for answer retrieval from unstructured data are Title-in-clue search, Full-text search, Document search, and Concept search.

Title-in-clue search looks for articles that contain a question clue in their titles and returns the first sentence of them. Full-text search strategy looks for the clue in the articles' title and text and returns the sentences from retrieved articles. Each sentence in the retrieved documents is assigned a weight based on the weight of the containing clues, and the top sentences are returned. Document search is performed over full text of the articles, and the top articles with their titles are considered as candidate answers. Concept search retrieves the articles whose title is the same as a question clue.

- **Answer analysis:** In the answer analysis step, multiple features are extracted from the answer text. The most important feature is matching the LAT of the question with LAT of the answer sentence. LAT of the answer sentence is recognized according to different strategies based on the method used for extracting the answer. For example, an answer including a number has *quantity* LAT, the focus of the answer is searched in WordNet, and the value of "instance of" relation is utilized for creating the LATs.

  Matching of the question and the answer LATs is performed using the WordNet hypernym relation and named WordNet specificity score. WordNet specificity score is calculated based on the traversals required for moving from answer LAT to question LAT. WordNet specificity score decreases by increasing the number of needed traversals.
- **Answer merging and scoring:** A simple version of the methods used for merging and scoring in the DeepQA pipeline is used here. The merging is performed by merging the candidate answers that are extracted from a common text. Features of the merged answers are determined by applying max-pooling

over each feature. The features happening in less than 1% of the questions and 0.1% of the answers are rare features, and they are eliminated. Additional features are created for logical features to explain them.

- **Successive refining:** Some extra scoring and refining steps can be executed in this pipeline. The first 25 best answers ranked by the initial scoring functions are selected and passed to later scoring functions. The last scoring function selects one answer among them as the final answer of the system.

8.4 Summary

We devoted this chapter to studying real applications of QA systems. We provided a detailed explanation of the overall pipeline of three well-known QA systems including IBM Watson, DrQA, and YodaQA in this chapter. IBM Watson is based on the DeepQA project and is developed to compete in *Jeopardy!* challenge. IBM Watson utilizes both the textual and knowledge graph data sources for extracting the answer of the question. IBM Watson is explained in Sect. 2.3.2. DrQA is another QA system which relies only on textual documents to answer the input question. We devoted Sect. 8.2 to study the DrQA model. YodaQA is an open-domain QA system inspired by DeepQA architecture that is explained in Sect. 8.3.

References

Baudiš, P., & Šedivý, J. (2015). Modeling of the question answering task in the yodaqa system. In *International Conference of the Cross-Language Evaluation Forum for European Languages* (pp. 222–228). Springer.

Berant, J., Chou, A., Frostig, R., & Liang, P. (2013). Semantic parsing on Freebase from question-answer pairs. In *Proceedings of the 2013 Conference on Empirical Methods in Natural Language Processing*, Seattle, Washington, USA (pp. 1533–1544). Association for Computational Linguistics. https://www.aclweb.org/anthology/D13-1160.

Chen, D., Fisch, A., Weston, J., & Bordes, A. (2017). Reading Wikipedia to answer open-domain questions. In *Proceedings of the 55th Annual Meeting of the Association for Computational Linguistics (Volume 1: Long Papers)*, Vancouver, Canada (pp. 1870–1879). Association for Computational Linguistics. https://doi.org/10.18653/v1/P17-1171. https://www.aclweb.org/anthology/P17-1171.

Ferrucci, D., Brown, E., Chu-Carroll, J., Fan, J., Gondek, D., Kalyanpur, A. A., Lally, A., Murdock, J. W., Nyberg, E., Prager, J., et al. (2010). Building watson: An overview of the deepqa project. *AI Magazine, 31*(3), 59–79.

Miller, A., Fisch, A., Dodge, J., Karimi, A.-H., Bordes, A., & Weston, J. (2016). Key-value memory networks for directly reading documents. In *Proceedings of the 2016 Conference on Empirical Methods in Natural Language Processing*, Austin, Texas (pp. 1400–1409). Association for Computational Linguistics. https://doi.org/10.18653/v1/D16-1147. https://www.aclweb.org/anthology/D16-1147.

Prager, J., Chu-Carroll, J., Czuba, K., & Maybury, M. (2004). A multi-strategy, multi-question approach to question answering. *New Directions in Question-Answering*, 21.

Rajpurkar, P., Zhang, J., Lopyrev, K., & Liang, P. (2016). SQuAD: 100,000+ questions for machine comprehension of text. In *Proceedings of the 2016 Conference on Empirical Methods in Natural Language Processing*, Austin, Texas (pp. 2383–2392). Association for Computational Linguistics. https://doi.org/10.18653/v1/D16-1264. https://www.aclweb.org/anthology/D16-1264.

Schlaefer, N., Gieselmann, P., Schaaf, T., & Waibel, A. (2006). A pattern learning approach to question answering within the ephyra framework. In *International Conference on Text, Speech and Dialogue* (pp. 687–694). Springer.

Chapter 9
Future Directions of Question Answering

Abstract The main important systems that are mentioned in this book focus on QA systems that use textual data or knowledge bases. The systems are trained on a dataset and are used to answer questions from the same domain and language. Various extensions have also been proposed to build QA systems with more specific features. The recent studies on QA show that still there is room to provide systems with more abilities, e.g., cross-lingual QA systems, explainable QA systems, zero-shot transfer model for QA, and why-type QA. In this chapter, we will list different directions that can be taken into consideration for the future of QA. We also name some related works on these topics which show their importance and the early steps that have been taken so far to reach these goals.

9.1 Cross-Lingual QA

Although various models and systems have been proposed for English QA, there are limited systems for other languages, especially low-resource languages. The main reason is the lack of training data for such languages.

Contextualized representation models paved the way. Cross-lingual training of language models for many languages, including Multilingual BERT (Devlin et al., 2019), XLM-R (Conneau et al., 2020), and Multilingual T5 (Xue et al., 2021), provides the possibility of using common vector space for various languages. In this case, a model can be trained on English data and be used for non-English languages with the help of transfer learning.

Considering the importance of this topic in QA, recent studies focus on developing cross-lingual datasets. XQA (Liu et al., 2019) is one of the cross-lingual datasets from Wikipedia which includes 28k samples and covers 9 different languages. TyDi (Clark et al., 2020) is another dataset in this family with 204k samples covering 11 languages. XOR QA (Asai et al., 2021) is built by re-annotating 40k samples from the TyDi dataset while covering 7 languages out of 11 languages.

S. Momtazi, Z. Abbasiantaeb, *Question Answering over Text and Knowledge Base*,
https://doi.org/10.1007/978-3-031-16552-8_9

9.2 Explainable QA

Using the state-of-the-art artificial intelligence techniques and mainly neural models achieved promising results in question answering. The neural models and deep learning approaches, however, provide no explanation about the output. The lack of explainability in the current methods motivated researchers to study a new area of QA, namely, explainable QA. The main ability of an explainable QA system is using an explainable computational model and an explainable interface that can provide the answer along with its explanation (Shekarpour et al., 2020). Different issues should be considered in such systems such as:

- The reasons and the situation(s) that an explainable QA system is required,
- The type of representation that should be used to provide explanation for end user,
- Evaluation techniques and metrics for evaluation of explainable QA systems.

Although in recent years different works have been done on this topic, they are limited to specific domain or type of data, such as medical QA (Zhang et al., 2019), and there is still room for further research on this topic.

9.3 Zero-Shot Transfer Models in QA

In most of the approaches and techniques reviewed in this book, the QA system is used on the same domain that has been used for training the system. This type of training limits systems to be used in real applications and industry-scale usage (Chakravarti et al., 2020).

Zero-shot transfer learning is one of the solutions to overcome this problem. In such a learning scenario, the system is not trained on a specific domain of data. Unsupervised or self-supervised models are well-known approaches for this goal. The advent of pre-trained language models provides the possibility of proposing advanced models for zero-shot transfer learning which needs to be studied in the future.

9.4 Why-Type QA

Factoid questions are the most usual type of questions in QA systems. User questions are categorized in different types based on their initial word, namely, who, when, where, what, why, and how. Questions starting with who, when, where, and what are known as factoid questions, while why and how questions are known as non-factoid questions. While factoid questions are easier to process and respond to, non-factoid questions need more advanced techniques and are not studied widely.

Focusing on why-type questions and how-type questions are an open problem that can be taken into consideration in the future. Answers to such non-factoid questions are subjective and require more detailed information which varies from a single sentence to a document. Advanced NLP techniques including pragmatic, discourse analysis, and textual entailment are normally required to better understand and answer these questions. Although some studies have been done on why-type questions in the first decade of 2000, they received less attention in the last years. This indicates the current gap between the state-of-the-art models in factoid QA and non-factoid QA.

9.5 Advanced Evaluation for QA

Another important issue in QA is the evaluation metric. Although reported results in various systems show that they achieved a very good performance in the field, in real application, we still see large amounts of questions that cannot be answered accurately by such systems (Rodrigo & Peñas, 2017). This indicates that we need a better and fair evaluation scenario to better reflect the current status of QA systems. This issue is an open topic for further studies in the field.

References

Asai, A., Kasai, J., Clark, J., Lee, K., Choi, E., & Hajishirzi, H. (2021). XOR QA: Cross-lingual open-retrieval question answering. In *Proceedings of the 2021 Conference of the North American Chapter of the Association for Computational Linguistics: Human Language Technologies* (pp. 547–564), Online. Association for Computational Linguistics. https://doi.org/10.18653/v1/2021.naacl-main.46. https://aclanthology.org/2021.naacl-main.46.

Chakravarti, R., Ferritto, A., Iyer, B., Pan, L., Florian, R., Roukos, S., & Sil, A. (2020). Towards building a robust industry-scale question answering system. In *Proceedings of the 28th International Conference on Computational Linguistics: Industry Track* (pp. 90–101). International Committee on Computational Linguistics. https://doi.org/10.18653/v1/2020.coling-industry.9.

Clark, J. H., Choi, E., Collins, M., Garrette, D., Kwiatkowski, T., Nikolaev, V., & Palomaki, J. (2020). TyDi QA: A benchmark for information-seeking question answering in typologically diverse languages. *Transactions of the Association for Computational Linguistics, 8*, 454–470. ISSN:2307-387X. https://doi.org/10.1162/tacl_a_00317.

Conneau, A., Khandelwal, K., Goyal, N., Chaudhary, V., Wenzek, G., Guzmán, F., Grave, E., Ott, M., Zettlemoyer, L., & Stoyanov, V. (2020). Unsupervised cross-lingual representation learning at scale. In *Proceedings of the 58th Annual Meeting of the Association for Computational Linguistics* (pp. 8440–8451), Online. Association for Computational Linguistics. https://doi.org/10.18653/v1/2020.acl-main.747. https://aclanthology.org/2020.acl-main.747.

Devlin, J., Chang, M.-W., Lee, K., & Toutanova, K. (2019). Bert: Pre-training of deep bidirectional transformers for language understanding. In *NAACL-HLT*.

Liu, J., Lin, Y., Liu, Z., & Sun, M. (2019). XQA: A cross-lingual open-domain question answering dataset. In *Proceedings of the 57th Annual Meeting of the Association for Computational Linguistics*, Florence, Italy (pp. 2358–2368). Association for Computational Linguistics. https://doi.org/10.18653/v1/P19-1227. https://aclanthology.org/P19-1227.

Rodrigo, A., & Peñas, A. (2017). A study about the future evaluation of question-answering systems. *Knowledge-Based Systems, 137*, 83–93. ISSN:0950-7051. https://doi.org/10.1016/j.knosys.2017.09.015. https://www.sciencedirect.com/science/article/pii/S0950705117304161.

Shekarpour, S., Alshargi, F., & Shekarpour, M. (2020). Towards explainable question answering (xqa). In *Proceedings of the AAAI CEUR Workshop*.

Xue, L., Constant, N., Roberts, A., Kale, M., Al-Rfou, R., Siddhant, A., Barua, A., & Raffel, C. (2021). mT5: A massively multilingual pre-trained text-to-text transformer. In *Proceedings of the 2021 Conference of the North American Chapter of the Association for Computational Linguistics: Human Language Technologies* (pp. 483–498), Online. Association for Computational Linguistics. https://doi.org/10.18653/v1/2021.naacl-main.41. https://aclanthology.org/2021.naacl-main.41.

Zhang, Y., Qian, S., Fang, Q., & Xu, C. (2019). Multi-modal knowledge-aware hierarchical attention network for explainable medical question answering. In *Proceedings of the 27th ACM International Conference on Multimedia* (pp. 1089–1097). Association for Computing Machinery. ISBN:978-1-4503-6889-6. https://doi.org/10.1145/3343031.3351033.

Printed in the United States
by Baker & Taylor Publisher Services